AF778386

A Rural Carpenter's World

The American Land and Life Series

EDITED BY WAYNE FRANKLIN

WAYNE FRANKLIN

A Rural Carpenter's World

The Craft in a Nineteenth-Century New York Township

UNIVERSITY OF IOWA PRESS IOWA CITY

University of Iowa Press,
Iowa City 52242
Copyright © 1990 by
the University of Iowa
All rights reserved
Printed in the
United States of America
First edition, 1990

Design by Richard Hendel

No part of this book may be reproduced or utilized in any form or by any means, electronic or mechanical, including photocopying and recording, without permission in writing from the publisher.

Printed on acid-free paper

Library of Congress
Cataloging-in-Publication Data

Franklin, Wayne.
A rural carpenter's world: the craft in a nineteenth-century New York township/ by Wayne Franklin.—1st ed.
p. cm.—(American land and life series)
Includes bibliographical references.
ISBN 0-87745-277-6 (alk. paper)
1. Carpentry—New York (State)—Westford. 2. Carpenters—New York (State)—Westford. I. Title. II. Series.
TH5608.7.F73 1990 89-20613
694'.09747'09034—dc20 CIP

Figures 5, 16, 21, 23, and 42 courtesy of the New York State Historical Association. Photos for figures 3, 4, 11, 12, 15, 18, 35, 37, 38, 40, and 43 supplied by NYSHA. Photos for figures 13, 17, 19, 22, 24, 44, 45, and 46, from the Raymond C. Tyler Scrapbook, NYSHA Collections, supplied by NYSHA. All other figures courtesy of the author.

Frontispiece:
Abandoned graveyard north of Westford, New York

Contents

Acknowledgments

A carpenter who builds a house owes debts beyond the obvious ones. In writing this book about carpentry, I too found the debts accumulating with surprising speed. The help and encouragement given me by the staff of the New York State Historical Association in Cooperstown, especially by librarian Amy Barnum and her staff, and by Wendell Tripp, editor of *New York History*, were invaluable from the start. The collections in NYSHA's wonderful quarters—collections of manuscripts but also of tools, so spacious is the sense of history there—were essential to all I did.

Lewis J. Holmes, the great-grandnephew of "my" carpenter and himself the town historian in Westford, was a gracious and illuminating guide to the landscape where he had spent his life preserving the past. Even in a state blessed with so many official local historians, it seemed to me that Westford was indeed very fortunate in Lewis Holmes.

Mrs. John Ahern generously let me see the inside of James C. Holmes's old house, much as John Silvernail (whose grandparents had owned it at one time) spent an afternoon telling me about the house and the barn which once stood across the road from it.

During my tenure as a faculty scholar at the University of Iowa (1982–1984), the office of the vice-president for academic affairs, especially Fredrick Woodard, gave me generous financial support. Without that aid, the project might never have been begun, let alone finished. My colleagues in the American Studies program, who have always encouraged my work in the field of material culture and vernacular architecture, make it a pleasure to undertake impossible projects like this one.

Don Scheese read and commented on an early version of my work on James Holmes. Kathy Zawicki prepared the index for the final version.

Nan Wolverton was the first person to read this book, but that is the least of what she has given me.

Preface

By John R. Stilgoe

Along the board glides the jack plane again and again, shavings curling from its blade, smoothness taking form beneath its heel. About the board, and about the structure in which the board is a single, handcrafted component, contemporary scholars know a great deal. Artifacts deserve and receive close scrutiny nowadays not only from the educated general public but from historians, "aboveground archaeologists," preservationists, and other professionals determined to make mute objects speak. But *A Rural Carpenter's World* probes far more deeply. Professor Franklin has picked up one shaving from the shop floor and used it as a sort of prism, glimpsing not only the board on which the plane worked but also the tree from which the board came—and most significantly, the hand, eye, and mind that guided the plane shaping the board from rough lumber.

A Rural Carpenter's World demonstrates what scholars fascinated with artifacts, and particularly with structures, have long suspected. Most buildings, and almost all rural buildings, are anonymous today only because their written records vanished long ago. And their builders, now unknown as individuals and casually dismissed as "local carpenters," lack biography because too few bequeathed to posterity any sort of documentation of their lives other than planed boards fitted together in the houses, poultry sheds, barns, and other objects still extant everywhere in the Republic. Anonymity turns out to be surprisingly comfortable. Few scholars dig very hard into written material to learn much about an anonymous house. Instead they prefer the precise techniques of artifact analysis, the measuring of rooms and bays, the scraping of old paint layers, the sketching of fenestration, all revealing much about the structure to be sure, but rarely illuminating the shadowy artisans who built it. *A Rural Carpenter's World* passes into another realm of research, one vastly richer albeit one far more complex.

From a carpenter's diary, the rough-and-ready accounts-and-memoranda book kept by a rural New York craftsman in 1869, Professor Franklin probes the structure of a microcosm, the small world of a farmer-carpenter. Using census schedules, newspaper accounts, maps, and still-extant structures, he stitches together fragments of information into a coherent, fascinating whole. His book places the carpenter in context, demonstrating what carpentry meant in a hill county bypassed by the Erie Canal, hopeful of railroad access, worried about westward migration, and recovering from the Civil War. *A Rural Carpenter's World* demonstrates the reaches of carpentry not only by following the work of one man but also by comparing his work with that of other carpenters in the county and with that of other tradesmen. And the book probes too at what carpentry reveals of a place in time, using the work of one carpenter as a portal through which something of rural New York life is traced.

A sort of serendipity graces this work. Over the many years Professor Franklin assembled the bits of information always so scattered in a rural county, he tried pattern after pattern. In time he made connections less painstaking scholars never discover. *A Rural Carpenter's World* clarifies the place of transient carpenters in rural New York, for example, analyzing the origin and impact of men moving west, moving from one hoped-for building boom to another, from carpentry into some other trade, at times competing with established local builders, at times working for them. Such is the fineness of this study. The reader learning of changes in a place learns also of one carpenter's maturation toward the craft of cabinetmaking, and learns how intertwined are the two developments.

Artifacts speak only in truncated translation. *A Rural Carpenter's World* lifts their grunts and monosyllabic words into full sentences, rewarding its reader with a vastly more subtle view of rural building and providing a model for research elsewhere.

Introduction: Documenting the Crafts

Sometime late in 1868, probably when he was spending a few days in Albany, his state's capital, the New York farmer and carpenter James C. Holmes bought himself a new pocket diary for 1869 (fig. 1).[1] In this modest blank book over the next twelve months Holmes was to record, usually with the laconic tone often audible in such rural documents, what he did, what he witnessed, and what happened to him. He was privy to few important events. The one to which he might have made reference—the new Erie war being fought for control of the Albany and Susquehanna Railroad (later the Delaware and Hudson)—was duly reported in the local press. But in fact Holmes wholly ignored it, even as he blithely rode the trains whenever necessary.[2] Nevertheless, for Holmes 1869 was personally a very eventful year. Most importantly, both his own father, Calvin Holmes, and his father-in-law, David Shelland, died in March. Beyond the immediate grief these deaths caused, they led to the breaking up of two households and forced adjustments on others. There was Holmes's stepmother, Anna Howe Holmes, to find a new place for; and by year's end, there was Marinda Shelland, the mother of his wife Mary, to take into the Holmes house.

But every parent dies, and every death causes some readjustments among the living. What gives the 1869 diary of James C. Holmes its special interest is the fact that in that year Holmes was very busy as a carpenter and therefore used his diary as a means of keeping track of his trade. Such day-by-day records of craft activity in fact are rather rare.[3] Much more common are account books, in which jobs are entered

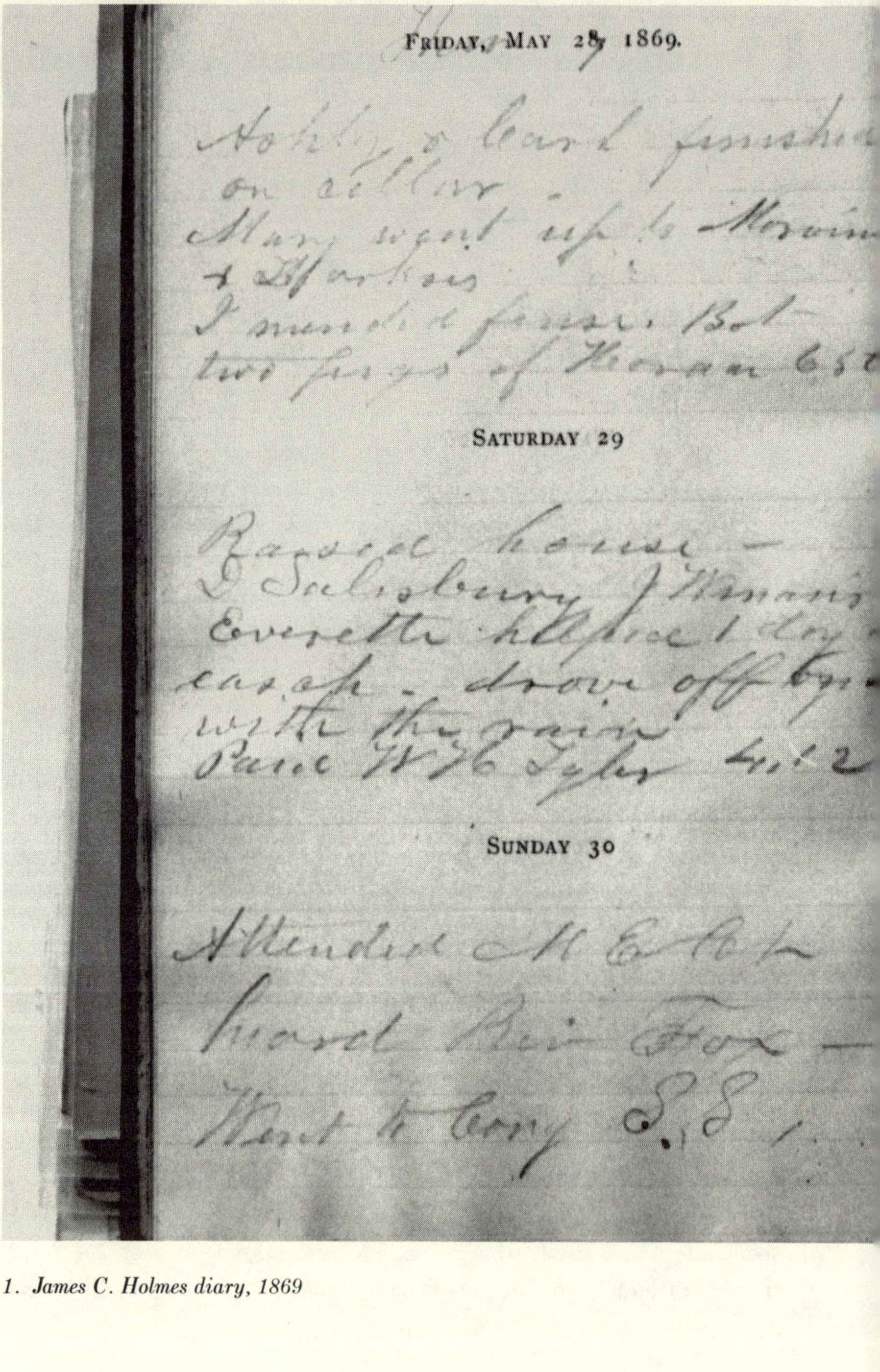

1. *James C. Holmes diary, 1869*

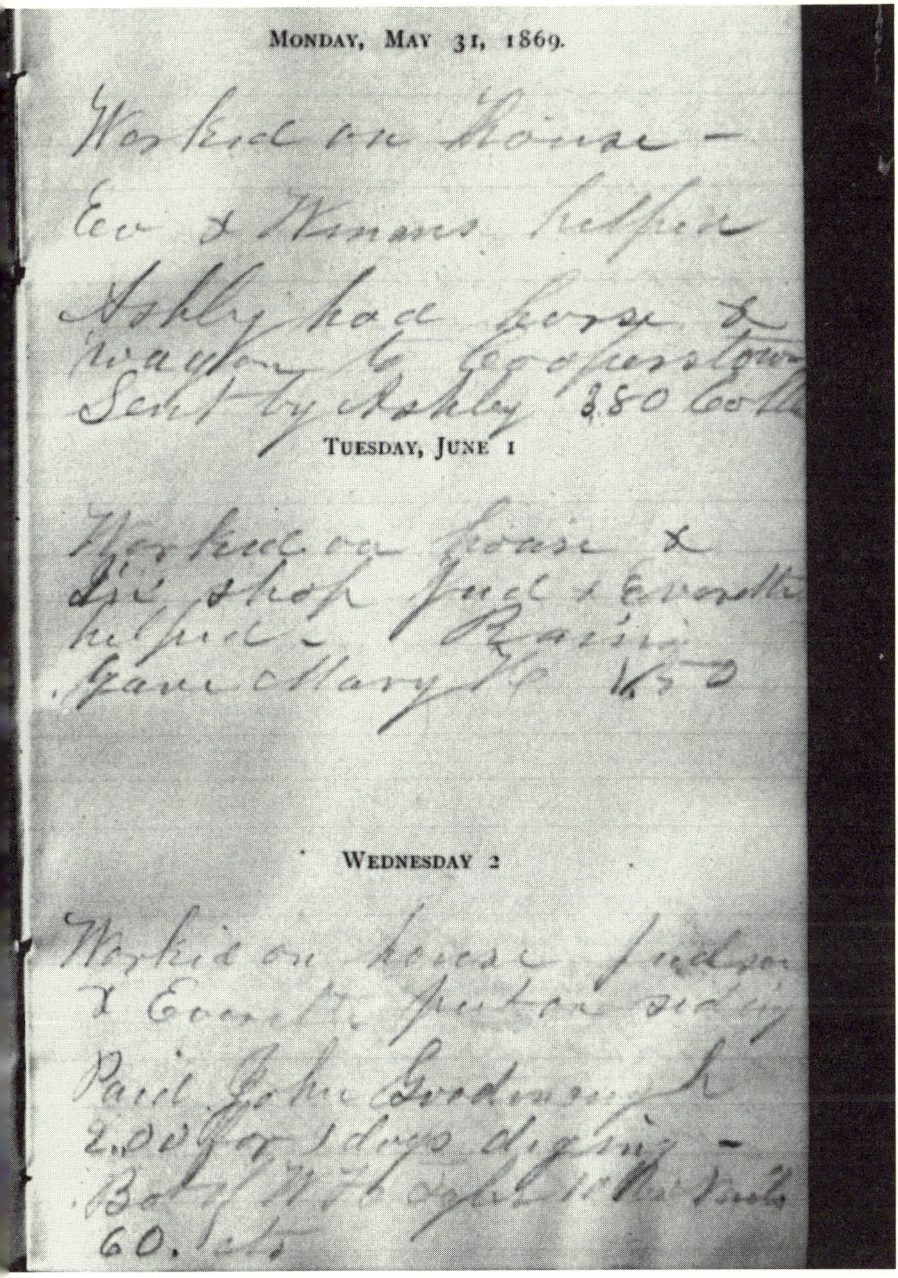

MONDAY, MAY 31, 1869.

Worked on House —
Ed & Simons helped
Ashley had horse &
wagon to Cooperstown
Sent by Ashley 3.80 Coll[illegible]

TUESDAY, JUNE 1

Worked on house &
ice shop Judd & Everett
helped — Raising
Gave Mary T[illegible] 1.50

WEDNESDAY 2

Worked on house Judson
& Everett put on siding
Paid John Goodenough
2.00 for 1 days digging —
Bot of [illegible]
60. cts

as simple line items, without any of the sense of process that, on the contrary, a diary almost by definition creates. Since Holmes and his workers were involved in building or significantly modifying six houses in 1869, his diary can give us an unusually concrete picture of the rural craftsman's procedures at a time when the coming industrialization of many aspects of the building trades had not yet had much effect outside American cities. It is primarily for this reason that the diary is reproduced here as the intensive centerpiece of a study of rural carpentry in Holmes's place and time.

I hope that the careful annotations of the diary text and its elucidation in the rest of the study will help place this craft document as complexly as possible in its historical and cultural context. With a couple of minor exceptions, each of the more than one hundred mostly obscure individuals referred to in the diary has been identified, and an effort has been made to recreate by these means the various social networks within which a craftsperson of Holmes's place and time worked. If nothing else, this study should disabuse us of some of the less defensible assumptions about the anonymity of the various crafts in the previous two centuries. To use that supposed trait as an excuse for overlooking documents and focusing exclusively on artifacts is at last to privilege one kind of evidence over another. And it is only by amassing as much evidence as possible, of as many kinds as possible, that we ourselves shall craft a better truth. I have no wish to deny that in any analysis of a craft the crafted objects must occupy a prominent position. But even if we have a wealth of objects at our disposal, we certainly are not justified in ignoring the perhaps less glamorous but no less useful documents that cluster around the crafts in the historical period. Certainly truth is not a consequence of one's methodological preferences; it is built by hard work and clear thinking. It is my hope that this case study will serve as an invitation to other scholars to devote serious attention to the documents of the American crafts.

But James C. Holmes and his diary for 1869 are merely the innermost circle in a concentric series here. By looking beyond his career to the place that carpentry held in his region not only in the 1860s but across much of the nineteenth century, I have tried to paint a much more inclusive portrait, a group biography of all the carpenters who practiced in Holmes's area up to 1900. Since that area was suffering from profound economic and social dislocations, I have been interested at last in the relative importance that men of Holmes's time placed on their skill. It has seemed to me that understanding their relation to modernization involves knowing whether they were more attached to a given place than to their occupations. I believe it can be inferred that in Holmes's case *where* he was mattered more than *what* he was. To that extent he stands before us here as a prime example of the rural American's refusal to make decisions based first and foremost on economic grounds. Something held him closely to Westford; if I have hinted at what it was—for he did not articulate it in the diary or in any of the other documents he left behind—I will count this project a success.

The diary proper forms chapter 2 here. The first chapter sets the immediate context for it; chapter 3 discusses the jobs Holmes undertook in the year 1869. The second part of the study, composed of chapters 4 and 5, attempts to set the diary and my discussion of it into a much broader framework. The fourth chapter thus portrays how a carpenter might fare in Holmes's region in the post–Civil War period, while the fifth presents a collective portrait of the sixty-one carpenters, including Holmes, who worked in his home township across the nineteenth century. I have been interested in the unyielding particulars of Holmes's craft and life; but I also have attempted, especially in these two last chapters, to gauge how representative Holmes was. In two appendixes, I have provided sketches of the people mentioned in the Holmes diary for 1869 and of the sixty-one Westford carpenters who are discussed en masse in the fifth chapter. The uses of biography are many. One use it has here is to remind us that, however tidy statistics are, the real lives which they manipulate resist mere summation. Indeed, they also resist language. No one knows better than I how much information I have failed to gather and how much surely remains to be gathered for much of what I discuss. But, like many Westford carpenters, I have other places to go, other jobs to put my hand to. At some point one must say, "Enough!"

PART ONE CRAFT

1 The World of James Holmes

James Calvin Holmes was born in the Town of Westford, Otsego County, New York, on May 25, 1830 (figs. 2 and 3).[1] Otsego County (fig. 4), which an 1860 gazetteer described as "hilly upland, divided into several ridges separated by deep, broad valleys,"[2] lies in the south central part of New York. Arrowhead-shaped, its point aims down toward the city of Binghamton and the Pennsylvania border. The two converging sides of the arrowhead roughly follow streambeds: Schenevus Creek and then the Susquehanna River on the eastern flank and the Unadilla River on the western. At the southernmost point of the county, the arrowhead's tip, the Susquehanna and the Unadilla join. The county's jagged upper line generally follows the rocky highlands that divide the watershed of these streams from that of the Mohawk. From many vantage points along this upper line, wonderful views of a gentler landscape to the north charm the eye. In the middle distance, the Mohawk runs peacefully in its broad, fertile valley, with the distant Adirondacks lost in the haze beyond. At one's back, Otsego thrusts its crest upward like a piece of Pennsylvania forced north.

In the contrast of these topographies lies much of the story that is to be told here. For near the Mohawk, with its many large factories and its dense population of urban workers, the legendary Erie Canal once ran. On that artificial river floated vast quantities of freight east and west, freight that converted New York from a minor member of the new American union into the major American state. But the canal, for all its magic, also left large parts of New York in the backwaters of their

2. Looking southeast toward Westford, along the Middlefield road

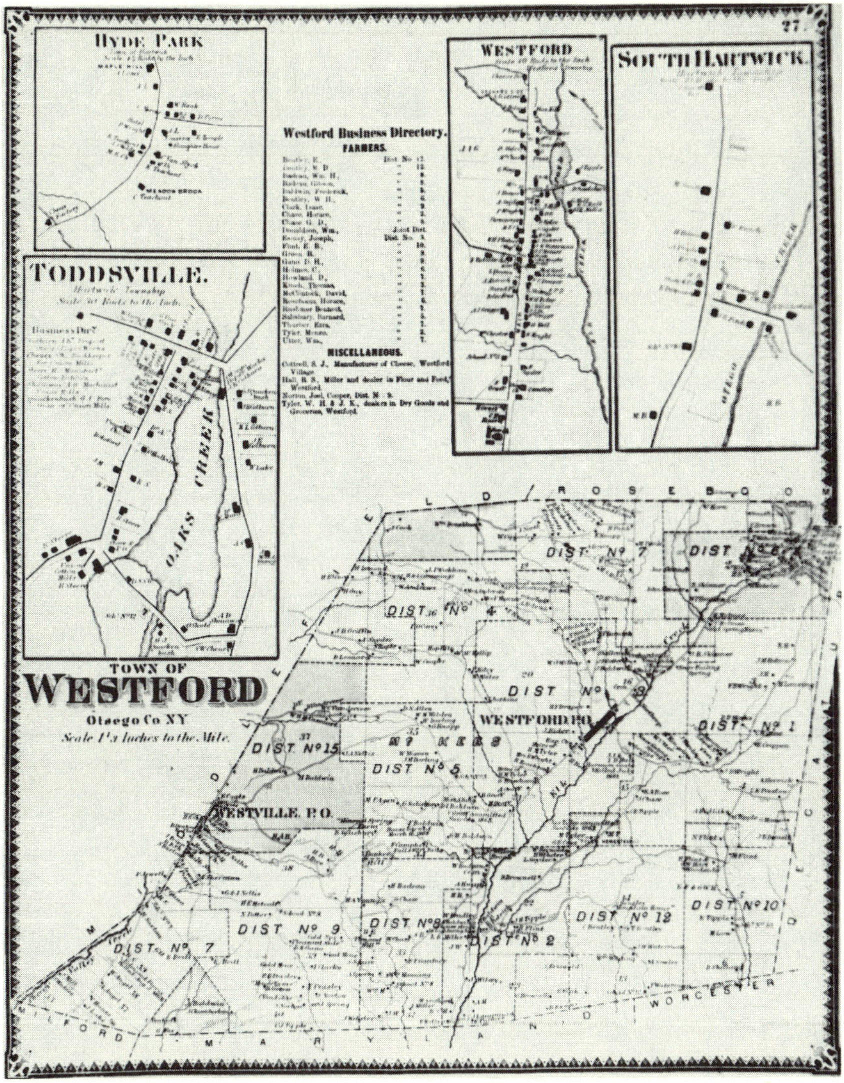

3. F. W. Beers, map of Westford township and Westford village, 1868

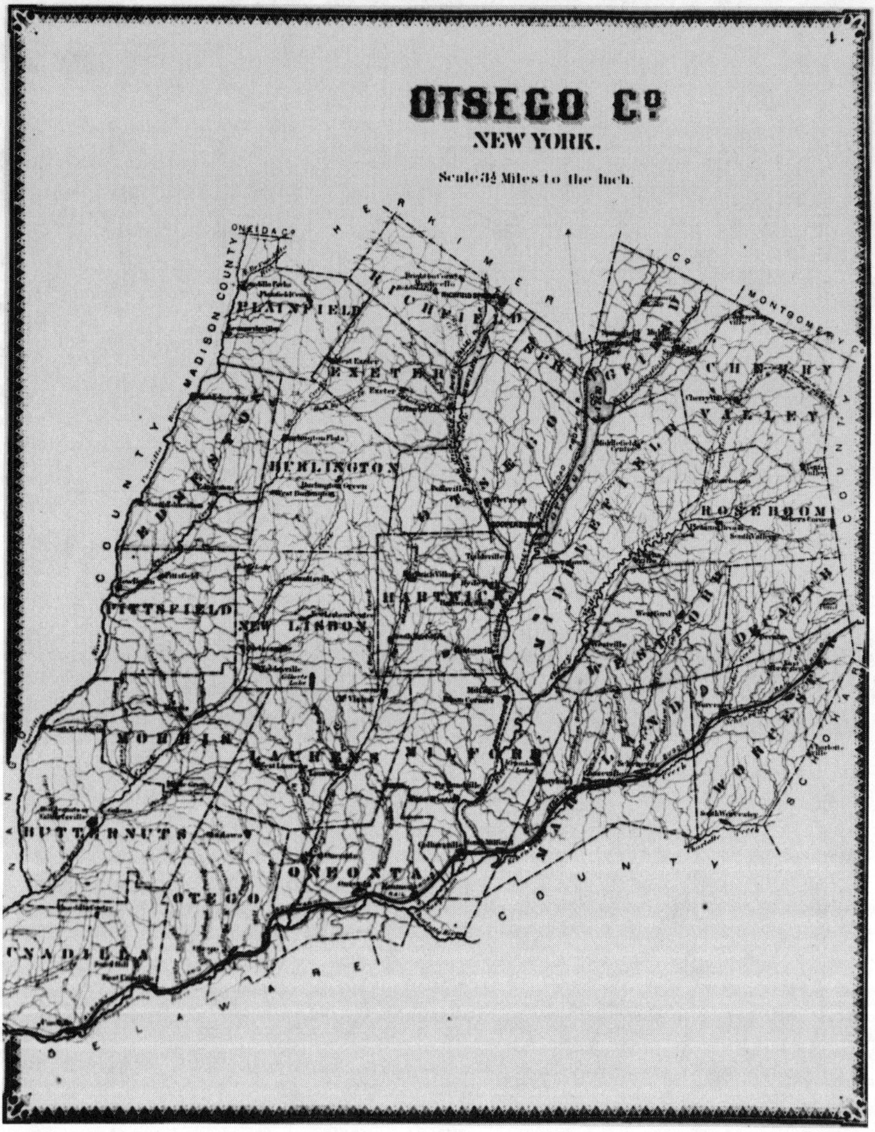

4. F. W. Beers, map of Otsego County, 1868

old condition. Indeed, in some ways it reversed the prosperity those regions had known prior to the canal's opening in 1825. Modernization entailed the displacement of some areas as others advanced. Such was the case with the Otsego region.

Set off from the vast domain of old Montgomery County in 1791, Otsego embraced in 1860 some 1,038 square miles. It was difficult terrain, with a good deal of rich soil but few ready means for shipping out produce. Like heavy corduroy, its dominant ridges ran down from the county's top toward the Susquehanna, severely limiting travel in the pre–Civil War period and retarding railroad development here much as a similar topography did throughout the larger "southern tier" region of which Otsego is one small part. But this landscape is lovely, hard as it can be. Watering the fertile intervales between the ridges are five major creeks that roughly parallel the course of the Unadilla: Butternuts, Otego, Cherry Valley, Elk, and Parker. At the upper end of the well-watered county lies the ten-mile long, blue-green expanse of Lake Otsego, the "Glimmerglass" of James Fenimore Cooper. On either hand as one looks up the lake, forested hills jostle the water; at the upper end, crosswise, lies the mountain known locally as the "Sleeping Lion." From the lake's foot in Cooperstown, the county seat, the Susquehanna takes its rise, cutting south toward Pennsylvania. Passing through the center of that state, it eventually empties into the Chesapeake.

At the county's northeast corner lies its oldest area, Cherry Valley. First settled in 1740, when it was a borderland of the sparsely settled Mohawk valley, this settlement became the scene of a bloody massacre during the Revolution. Resettled after the war, the village gave its name to one of the two large townships into which the new county of Otsego first was divided. Cherry Valley itself was progressively subdivided over the years, until it was merely one of Otsego's twenty-four townships by the time Holmes kept his diary. The area that was to become Westford remained part of Cherry Valley until 1797, when the large southeastern segment of the latter became Worcester township. That township in turn was divided in 1808 into four parts—Westford and Decatur on the northwest and northeast, and Maryland and Worcester, respectively, below them.

Westford, embracing some twenty thousand acres, lies athwart the bed of Elk Creek. As one travels up Elk Creek road from the village of Schenevus, the main settlement in Maryland township, the ridges on either side move in closer and closer. Finally the narrow vee closes on Holmes's village of Westford, located where Elk Creek tumbles over the waterfall that first attracted settlers to the site. This trip along Elk Creek was one Holmes made several times in 1869 as he carried materials for his various jobs back from Schenevus, his main trading center.

Westford village was already occupied in the 1790s, a mill having been built on the stream there as early as 1794. The nascent village was important enough that from 1797 until 1808 the winter sessions of the Worcester town meeting were held there, and from 1808 on it was to remain the largest gathering of population in the new, rather thinly settled township that was named for it. Only one other settlement, Westville, has ever existed in the township, and half of Westville in fact lies in Middlefield township. Other informal places, however, were well enough discriminated among the local population to acquire names: Bentley Hollow—also called Maple Valley—and Skinnerville, both north of Westford village, are the two most important. This was hardly a densely populated area. Even Westville and Westford village were simply aggregative settlements, points where houses clustered close to one another along a through road. Westford, it is true, had somewhat more depth. In addition to its main axis—the Elk Creek road running northeast from Schenevus—a staggered crossroad came in from Middlefield village to the west and went out, from a point slightly farther north, toward the village of Worcester to the southeast. Especially in the latter case, a number of structures (including the gristmill and sawmill operating in 1869) had been built along these side roads. It was on the Middlefield road that James Holmes had lived for a number of years following his first marriage; it was on the Worcester road that he built himself a new house in the year he kept the diary printed below.

As was true with most of Otsego's townships, the major economic activity in Westford throughout Holmes's life was farming. In 1865 some 15,000, or three-fourths, of Westford's acres were "improved," with a total farm value of $1.67 million and a livestock value of

$113,260. The township produced a mixed group of crops and animals. On the 3,372 acres plowed for planting in 1865 (5,664 were in pasture, 4,775 in meadow), Westford farmers raised winter and spring wheat, winter rye, oats, barley, buckwheat, corn, potatoes, peas, beans, turnips, flax, and hops. The major crops were oats (1,702 acres), hops (352¾ acres), corn (347 acres), buckwheat (344 acres), potatoes (208⅜ acres), and flax (57¾ acres, the county's high by a factor of more than 2). There were some 12,000 apple trees, a maple sugar production of 271 pounds, and a honey production of 1,504 pounds (1864's figure). Milch cows kept in the township numbered 923; from them in 1864, 102,250 pounds of butter had been produced (in the middle range for the county, whose high was 221,815), as well as a modest 5,841 pounds of cheese (here the county high was a whopping 569,659 pounds). Pork production was low (80,963 pounds sold in 1864), but sheep numbers (4,278) and wool production (15,735 pounds) were both high by Otsego County standards. These figures suggest that Westford, though it possessed good soil in its arable districts, was largely devoted to cattle and sheep production, both of which activities could put the township's many hilly areas to good use.

James Holmes's parents were farmers, as were most of his siblings. By the time he was twenty, however, the federal census shows that Holmes had moved off the land and was living as a "student of medicine" with Dr. Calvin Howard in Delhi, Delaware County. This apprenticeship seems to have led to no permanent changes in the young man's life. In 1855, five years later, he was back in Westford, just married to Mary Shelland and now declaring himself a carpenter. He was to repeat the same declaration at each census up to 1875, when he was a "farmer and carpenter," and 1880, when he was a "farmer" pure and simple. But it would be wrong to read these declarations—or the changes in them—too rigidly. The Holmeses were buying land throughout this period: a 5-acre farm and a village lot adjoining it in 1860, a 27-acre parcel in 1868 (soon sold but repurchased the next year), and three adjacent parcels, totaling 52 acres, also in 1868. In the 1870 census of agriculture, which contains an entry for "carpenter" Holmes, he reported how he had used this last acreage during the year ending June 1. The farm, all of it "improved" ground, was worth

$5,000. Holmes kept one horse, six milch cows, and another cow or calf or steer, all of them together worth $400. From the cows Holmes—in all likelihood this is evidence of Mary's labor—had produced 450 pounds of butter, which in 1869 would have brought in between $150 and $175. He had raised 20 bushels of Indian corn, 80 of oats, and 25 of Irish potatoes, had mown 6 tons of hay, and had produced 100 pounds of maple sugar. Holmes estimated the total value of this modest production at $315.00—no great amount, although in fact this was equivalent to what it would have taken Holmes twenty-one weeks as a carpenter, at $2.50 a day, to earn. Since much of the butter probably was sold (although the diary records the sale of only 209 pounds for a total of $68.30 in 1869), that commodity alone would have provided the equivalent of more than two months' wages.

Still, this was modest. When Holmes called himself a farmer as well as a carpenter in the state census five years later, the agricultural statistics showed that he had significantly expanded his farm. In 1874 Holmes had plowed 15 acres, had kept 5 in pasture, and had 25 in meadow; in 1875 (the census gives some figures for both years) he had plowed 12, increased the pasture to 8, and kept the same acreage in meadow. The hay harvest in 1874 had been 25 tons. The 5 acres of oats yielded 150 bushels in 1874, 5 acres of winter rye sown in 1873 produced 20 bushels, and from 4 acres of buckwheat 73 bushels had been harvested. His 50 apple trees (planted, as the diary records, late in 1869) had borne 10 bushels of fruit in 1874; this wasn't much, but then the trees were still shy of their full productive years. The total of 7 milch cows he had milked that year had produced 900 pounds of butter (twice the amount of 1869–1870) and from the 2 swine slaughtered then he had made 500 pounds of pork. In addition, he owned $5 worth of poultry and had sold $10 worth of eggs (this, again, is probably evidence of Mary's activity). His gross sales from the farm in 1874 totaled $400, well above the total value of everything he had raised (both for the market and for home use) in 1869–1870 and equivalent to perhaps thirty weeks of carpenter's wages.

In 1875, Holmes reported in June, he had planted 1 acre of corn, 1 acre of potatoes, and 10 acres of oats. His hay, of course, had not been mown yet but probably would have been about as productive as the year before. Earlier in the year, he had made 400 pounds of sugar and 15 gallons of maple molasses from his sugar maple trees. He now was milking 6 cows, had 4 bulls, 1 new heifer calf, 2 new pigs, and 2 horses that were older than two years.[3]

Comparing James Holmes's agricultural activities with those of his father, Calvin, in 1855, when Calvin was still quite active and was depending on farming as his sole support, suggests that in 1874 and 1875 James in fact was a serious farmer by local standards. In 1855, his father had owned 113 acres; 88 were improved and 25 were unimproved. He kept 35 acres in pasture, 35 in meadow, and out of the rest he used 10 for oats (150 bushels), 4 for rye (40 bushels), 1 for corn (20 bushels), 1 for potatoes (140 bushels), ½ for buckwheat (10 bushels), and ¼ for beans (5 bushels). With the exception of the last item, this array of crops was remarkably similar, in both type and quantity, to what James Holmes was planting two decades later, in the mid-1870s. Calvin Holmes in 1855 reported owning the following livestock: 7 milch cows, 3 butter cows (which produced 300 pounds), 3 horses, 6 swine, and 63 sheep (from which he produced 43 fleeces), as well as some poultry (he sold $6 worth of eggs). Here, too, there are differences, the most significant of them being the sheep. But on the whole, during his active career as a farmer, Calvin was operating within the same range as his son, James, did twenty years later. In 1875, we may conclude, James was deriving a good deal of his support from the farm. He also was practicing a kind of diversified agriculture, characterized by a variety of crops and animals in mostly modest acreages and numbers, that was historically typical of the region.

This last conclusion is borne out by a contrast with the census entries for his older brother Joseph Marvin Holmes. In 1855, when he had been married for nine years and was twenty-nine years of age, Joseph Marvin Holmes reported the following data: he had 60 improved acres and 20 unimproved, raised 200 bushels of oats on 10 acres, 75 bushels of corn on 2 acres, 150 bushels of potatoes on 1½ acres, 25 bushels of buckwheat on 1 acre, and 3 bushels of barley on ½ acre. He had picked 50 bushels of apples from his orchard, harvested 50 pounds of honey, made 300 pounds of maple sugar and 3 gallons of maple molasses, owned 4 neat cattle and 7 cows (from which he made 800 pounds of butter), 3 horses, and 5 swine. Ten years later, in 1865, Joseph Marvin Holmes reported owning much more land, 170 improved acres and 60 unimproved. For the most part, his acres under cultivation remained close to the levels of 1855. He was still raising 10 acres of oats (but harvested 300 bushels) in 1864, for instance, although the next year he cut back to 6 acres. Likewise, he had planted 3 acres of potatoes in 1864 (but only 1 in 1865), and had harvested 200 bushels. The big changes came in other areas. He now owned 20 milch cows, all of them butter-producing, and from this herd in 1864–1865 he had produced 2,000 pounds of butter. Only three Westford farmers topped this production in 1865. Secondarily, Joseph Holmes had begun to grow hops, a potentially very lucrative but always temperamental local crop. In 1864, he had a 2½-acre hopfield; this had produced 1,000 pounds. During 1865, he had expanded the field to 4 acres.

Five years later still, in 1870, the picture was similar. Joseph Holmes now reported owning 190 acres of improved land and 70 of woods. His production of hops had nearly doubled (to 1,800 pounds; prices in October of 1869 were about twenty cents per pound).[4] Although he still owned only 20 milch cows, they had yielded more in the year ending June 1, 1870, a total of 3,000 pounds of butter. The total value of his produce in that period was $2,765, or nine times as much as his younger brother James reported in the same census. Joseph was a very serious farmer by local standards. Only three farmers in the town of Westford that year reported a higher dollar figure. Only one owned more milch cows, and not even he reported making as much butter. The next highest figure was 2,500 pounds, which had been the high figure as recently as 1865 but which Joseph Holmes now surpassed by fully 20 percent. In fact, only 35 of Westford's 256 farmers reported making 1,000 or more pounds in 1869–1870.

The resulting picture is clear. Unlike his father, Calvin, or his brother James, Joseph Marvin Holmes had specialized. He also had responded increasingly to the trends of a modernized market agriculture, for butter and hops were two prime market crops in the region at that time. James, while still relying on his craft for income, had developed a smaller farm that provided some cash income but was not really an instrument of the market. In this he was closer to the lingering Westford norm. Like many of his neighbors, he lived a composite economic existence, deriving partial support from many activities. Even his craft, as the 1869 diary makes abundantly clear, was a composite of many different kinds of activity and production.

These contrasts are sharpened by a quick reference to the experience of James's younger brother, Horace. When Horace Holmes began running Calvin's farm in the 1860s and then bought out the other heirs in the 1870s, he followed Joseph's lead rather than the precedent of Calvin or James. Already in 1865, when Calvin was relying heavily on Horace, the farm's butter yield had risen from 300 pounds (the 1855 figure) to 1,000 pounds, and by 1870 this had risen further to 1,200 pounds. Interestingly, in 1869, it was not the older brother James who was supplying the younger one, Horace, with hogs or cows but just the reverse. One senses that there was, in this family, a decided split between the old and new ways. And James largely chose the old ones. Although it might be claimed that he had learned carpentry not out of a desire to escape from agriculture but rather as a means of amassing enough capital to enter it in earnest—as at last he had by 1880—"in earnest" in his case meant something decidedly different from what it meant for either Joseph or Horace.

James Holmes came from a large family and in 1869 was building his own, ultimately smaller, one. The members of both are profiled in appendix 1, as are all the local residents mentioned in the Holmes diary. Here it will be useful to say something about some of the more important craft and community connections that Holmes enjoyed. The kinship network within which James Holmes lived and worked in Westford surely supplied him with a relatively large and probably sustaining community. But that network was not the only important social force in Holmes's life. At various times, for instance, he held a variety of minor civil positions. He was elected town clerk for 1861, and in 1863, 1864, and 1866 he served as clerk of elections. He held the post of inspector of elections in 1865 and again in 1873 and 1876.

But Holmes's most important social ties came from another quarter. This was the local Baptist church. James Holmes and Mary Shelland were married at the Baptist parsonage in Westford in May of 1855 and later that year were baptized and joined the congregation. This congregation dated from 1825, when it had been founded as an offshoot of one in nearby Middlefield. Through much of its life, the Westford Baptist church had a tenuous existence. Often—as in 1869—it was without a minister, and when it had one it had difficulty paying him. Often, too, its membership declined to dangerously low levels, or when it remained high enough in numbers was low in spirit. But the church had a core of the active faithful. These included the family of Milton Wright, the man from whom James Holmes's brother John Holmes (and probably James himself, as well as a good many other local artisans) learned carpentry: in fact, James Holmes and Milton Wright at one time served together on the Baptist building committee, the body charged with maintaining and recommending repairs for the sanctuary. Wright's association with the Westford Baptists was especially strong. His father, Pomeroy Wright, had been one of the two founding deacons of the congregation in 1825, and Pomeroy remained a deacon until his death in 1876.

James Holmes's family also had wider ties with the congregation. Several of his siblings—including Lyman (who died elsewhere in 1850), Lucy, Joseph Marvin, and Clarissa—had joined the Baptist church before James did, although it is arguable that by 1855 only Clarissa was still active. James Holmes remained central to the church for the rest of his life. He was elected a trustee in 1858, and he succeeded Milton Wright in 1862 as church clerk (fig. 5), a position he held until early 1869. He was repeatedly chosen to fill various church positions—treasurer and collector, member of the pulpit committee, member of the building committee, and delegate to local Baptist association meetings and local ordinations. In the later 1870s, Holmes was founding superintendent of the sabbath school in Westford and shortly thereafter was elected president of the Worcester Baptist Association Sabbath School Convention. As late as 1892, Holmes was again elected treasurer of the Westford congregation.

And as late as March 2, 1895, Holmes read the congregation's covenant at its monthly meeting. The next day he attended church, had communion, and went dutifully to the sabbath school. Once back home, he went across the road to tend to his cows, only to collapse and die in his barn.[5] His fellow Baptists mourned him the way they tended to

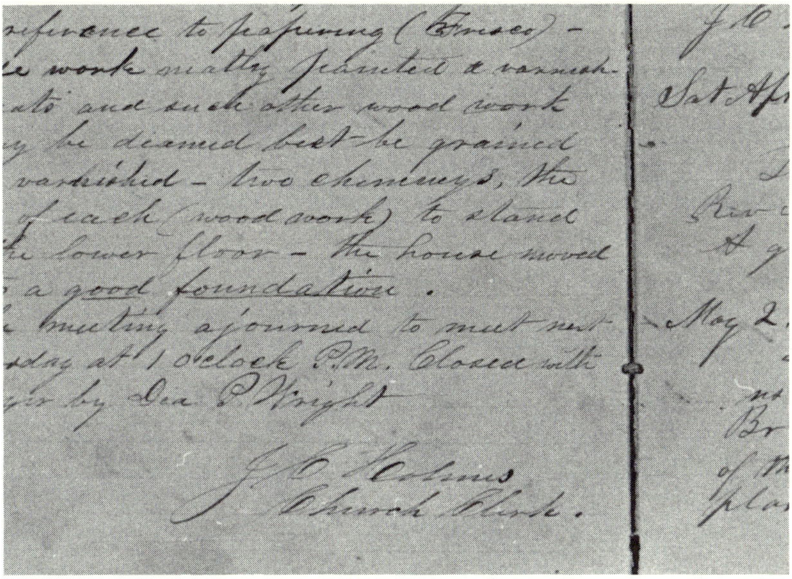
ference to papering (Frisco) -
e work neatly painted & varnish-
ato and such other wood work
y be deemed best - be grained
varnished - two chimneys, the
of each (wood work) to stand
he lower floor - the house moved
a good foundation.
meeting ajourned to meet next
day at 1 o'clock P.M. Closed with
er by Dea P. Wright
J. C. Holmes
Church Clerk.

5. *James C. Holmes's signature, Westford Baptist Book B, 1863*

mourn that handful of members who stood by the frail church through its worst times. The current clerk, J. L. Howland, wrote in the clerk's book that the sad event had left them all "in deep grief." "We feel," Howland continued, "that by his death we have lost one of our most faithful workers, and his voice will be sadly missed in all the appointments of the church."[6] He had, indeed, been one of the "faithful workers," not only in the church but in the community at large as well.

HOLMES'S HELPERS

It will be useful to speak briefly here about the dealings James Holmes had with other carpenters in Westford during 1869. The most important of these other men were Judson Winans (b. ca. 1839) and Everett Herdman (b. ca. 1852), both of whom worked with Holmes for considerable portions of the year. In 1855, when Winans first came to Westford with several siblings, he was a sixteen-year-old orphan. The Winans children probably had been living in Michigan when one and then the other of their parents, or perhaps both of them at once, died. They were brought to Westford to live with their aunt, Louisa Winans, and her husband, Holmes's associate Milton Wright. Judson Winans probably stayed with the Wrights for several years and probably learned carpentry from his uncle Milton. By 1860 he had married Louise Darling, a Westford woman, and was living with her and his sister, Marion, in his own house. Employed then as a laborer, he probably was working either in the building trades or as a fieldhand for someone else, since he owned no land of his own until 1862.

His purchase of sixty acres in that year may have marked some desire to enter agriculture in a serious way. But Winans turned around and sold that parcel again in 1863, and the fact that he realized almost 100 percent profit on the resale suggests that he had bought bare ground and erected a "spec" house on it. He likewise seems to have built "on spec" on other occasions. The house that Holmes helped him build late in 1868 and early in 1869 (this was the first job recorded in the diary, as we shall see) thus was sold off relatively soon, as Winans and his family moved to the booming city of Oneonta. In that city, near the lower border of the county, the coming of the A & S Railroad, and especially the railroad's shops, was giving a considerable stimulus to local construction. Between 1865 and 1875, almost half the increase in the number of houses reported in the state census for Otsego County was credited to the township of Oneonta alone.[7] This was a situation that Winans, a kind of nascent developer, seems to have found very attractive. He set himself up in Oneonta for the first time explicitly as a carpenter, his trade in both the 1870 census and the 1872 county directory.[8] He never returned to Westford.

Everett Herdman was a Westford native whom Holmes probably had known since Herdman's birth. In 1855, when Holmes and his wife were living in Chester Wright's house, Everett's parents, farmers John and Acenath Herdman, were living very close by. Everett was never credited in any Westford census with an occupation: only thirteen in 1865, by 1870 he had left the town, perhaps because he married, perhaps to find a job elsewhere. Relatively little is known about him as a result, and efforts to trace him after 1869 have proved futile.

Winans and Herdman served different functions for Holmes and had different rates of pay. Holmes hired Herdman in April for $18.00 per month. Since a month was composed of twenty-four ten-hour days, his wage was just $.75 per day. It is clear from various calculations Holmes performed that he himself was making $2.50 per day (see, for instance, his "House for Self" summary, memorandum 6). Judging from this sharp difference and from the kinds of tasks Herdman is recorded as performing in 1869, it is right to call Herdman Holmes's helper rather than a carpenter in his own right. Herdman was never set upon a carpentry task, for instance, without the supervision of someone more skilled than he—usually either Holmes or Winans, though occasionally others temporarily helping Holmes. Within these limits Herdman did have rather wide experience in the seven months Holmes employed him. He helped Holmes prepare framing timbers, sided the frames once they had been erected, and put down roof boards and then shingles. Inside, he installed flooring and put up partitions, helped Holmes case the windows and doors, and on one occasion (for Holmes's house in August) mixed mortar for the plasterers. But Herdman also played the part of a fill-in laborer, and his work extended well beyond mere construction. On his first day he helped Holmes move hay to a rented storage barn, draw two loads of firewood to Holmes's house, and clean away one load of manure. That afternoon, and for part of the next day, the two sawed firewood. Over the next months, although he mostly did construction work, Herdman also (and often without Holmes's help) was to perform a wide array of tasks. He planted potatoes, mended fences and dressed new rails, hoed Holmes's fields, mowed hay from Holmes's meadow (with Winans on one day and with one of Holmes's neighbors, and then Holmes himself, the next week), helped Winans dig Holmes's potatoes in October (these were the ones he had helped Holmes plant in May), and then spent one of his last days husking Holmes's corn.

Throughout the year, of course, Holmes himself performed a variety of tasks and seems to have expected others to be just as adaptable. Appropriately, the form of compensation that bound Holmes with his workers and his clients also was mixed. Although Herdman's wages were stated in cash and by the month, Holmes paid the young man neither by the month nor wholly in cash. Six payments made before

Holmes settled with Herdman in the fall (the payments began in late May and ended in mid-September) fall into no time pattern whatsoever, and the amounts involved, ranging from $1 on August 18 to $10 on September 11, also betray no real design. Between his start on April 13 and August 17, Herdman had worked a total of more than one hundred days for Holmes (memorandum 5) and hence was due more than $75. But even if the $1 paid him on August 18 is included, he had been paid by this time only $11. This means that Holmes's hired hand in effect was financing the operations of his employer; the $64 he had yet to receive marked an easing of the considerable financial pressures on Holmes in this very active year. The same was true in a more complex way for Judson Winans.

Holmes finally did pay off Herdman, but again no regular pattern is to be detected in their dealings. Once the young man stopped working for him in October, the two met to settle up. Holmes at this time paid Herdman $20.25, and they agreed that another $67.00 still was owed. Within a few days, Holmes paid Herdman $20.00 more and "sold him a cow [worth] $40.00." This last exchange was without question part of the settlement: Holmes had not so much sold Herdman the cow as used the cow as an in-kind payment. The account between the two was not cleared until more than two months later, when on December 24 Holmes made Herdman a final cash payment of $7. Why Holmes waited so long to finish his business with the young man is hard to figure, but how their relationship ended was in keeping with how it had run its course. Within broad agreed-upon economic limits, it seems to have been less a business relationship than a more nearly personal one. Cash was only one nexus between them.

That point holds with significantly more force in the case of Judson Winans, with whom Holmes had more dealings and more kinds of dealings, both in 1869 and before and after that year. Holmes must have known Winans since the young man's arrival in Westford in 1855. Judson was living then with the Wrights, very near the Holmeses. When Winans joined the Baptist church in 1857, Holmes and he became more than neighbors. Indeed, since it is likely that the several Winanses who were active in the Westford Baptist church before 1850 were Judson's relatives, young Holmes must have been familiar with the family years before Judson came—or perhaps came back—to Westford in 1855.

And there was the influence here, as in so many other places in Holmes's life, of the carpenter Milton Wright. Winans probably worked for his uncle, perhaps on jobs on which Holmes also was engaged, in the late 1850s and early 1860s. At the very least, he must have learned the rudiments of the craft from Wright. That he had learned a good deal from someone prior to 1869 is clear from the sorts of tasks he performed then for Holmes. Like Herdman, he helped Holmes on a number of different jobs, and on those jobs he too did a number of different things. His skill, however, was considerably greater than Herdman's. Herdman did very little shopwork beyond helping Holmes prepare large framing members, for instance, whereas Winans probably helped Holmes make trim for Holmes's new house in the shop during mid-March. For his own new house, Winans probably did at least some of the finish work that Holmes did not note doing himself in the first two months of the year. We know from the diary, furthermore, that Winans was singled out, presumably because of his skill, to undertake certain tasks alone or with Herdman as his helper. The two thus put siding on the Holmes frame together on June 2 (Holmes joined them in the task on June 3), and Winans painted the whole exterior of the house alone a few days later. Winans likewise lathed the interior on June 30, while Herdman and Holmes were busy about other things. Winans put the water table on William O. Ashley's new house by himself on August 30.

More instructive are Holmes's indications in the diary that Winans was beginning to assume supervisory control of whole jobs on several occasions. On June 14, Holmes spent at least four hours, presumably in his new shop across the road from the site of his partially finished dwelling, readying window frames for Myron Hubbard's house. Meanwhile, Winans and Herdman were inside the Holmes house laying the floor. On July 21, Winans was alone all afternoon at the Hubbard job, probably shingling or siding the frame, while Herdman and Holmes were tending to some farm work. On August 4, when Holmes put in only three-fourths of a day at Hubbard's, Winans and Herdman put in a full day. More importantly, on later jobs Winans was given essentially full charge. Holmes remained on the Ashley job in early September while Winans, with Herdman as his helper, began a new job—apparently an alteration rather than a full new house—for German

Baldwin about a mile away to the north. Presumably, Baldwin himself was then at work on the job, too, although there is no reason to believe, from what is otherwise known about his skills, that Baldwin knew even as much as Winans about carpentry, let alone more. Baldwin was basically a laborer; when he left his own job on October 11 and 12 to work for Holmes, he spent those two days, for instance, digging Holmes's new well.

The Baldwin job, whose nature is otherwise unknown, must have entailed a good deal of labor on Winans's part. From September 6 to October 13, excluding whatever contribution Baldwin (or his son Augustus) may have made to the effort, Winans and Herdman put in a total of almost six hundred hours. And during all this time, Holmes, although he may have visited the site, did not do any work at it; this was Winans's responsibility. In a minor way, the Skinner job, which lasted for five days in mid-October, confirmed and perhaps even extended Winans's skill. Here, because Herdman had stopped working for Holmes already, Winans was completely on his own.

Holmes nowhere specified the rate he was to pay Winans. Careful calculations based on their total economic relationship suggest strongly, however, that Winans worked for a wage twice that of Herdman: $1.50 per day or $36.00 per month. Although it isn't necessary to repeat here step-by-step the process by which this conclusion has been reached, a brief summary of the kinds of information used will indicate something of the range and complexity of the dealings between the two men. First, it is likely that Holmes nowhere indicated what he was going to pay Winans because in fact he hadn't really "hired" him as he had Herdman. That is, for a good part of 1869 Winans was laboring for Holmes in order to pay Holmes back for the time Holmes had spent on the new Winans house. Much as Holmes had paid Herdman off by means of some cash and a cow, so Winans was paying Holmes off in kind (by year's end, in fact, he had more than worked off his debt). This was a means, essentially, of creating the assets by which he was to discharge his liabilities. That Holmes tolerated such an arrangement suggests how entangled in their personal relationship the economic dealings between the two were. Although it seems likely that Holmes knew before the start of 1869 that he was going to build new houses for several clients besides Winans, and thus knew that Winans would be able to make good on the debt, it was Holmes's ability to secure those jobs that essentially allowed Winans to hire Holmes for the Winans job. So far as the records show, Winans paid Holmes only $16.90, this for shingles and lumber on January 21, toward the costs of building his own new house.

Winans worked about 125 days for Holmes in 1869 (memorandum 5). Holmes, on the other hand, worked 26 days on the Winans house in 1868 and 1869. These were the two major items in their dealings. But Winans was in Holmes's debt for a total of more than three thousand feet of hemlock lumber worth in the neighborhood of $50 (memorandum 1), plus shingles and other quantities of lumber. Winans, in turn, supplied Holmes with small amounts of lumber (on one occasion they swapped hemlock for flooring) and 18½ pounds of nails. Beyond the exchange of time between them, in other words, there was an exchange of goods, and the pattern of payments Holmes made to Winans (he made six from August to November, for a total of $63.70; Winans borrowed a total of $17.10 on two occasions) makes sense only if the cash supplemented the exchange in labor and materials. Holmes noted on November 20 that his final payment to Winans of $5.70 balanced accounts between them. Were the total of cash payments to Winans ($80.80) the only means of settling, Winans would have received a net daily wage of only $.65, less than Herdman's meager $.75. And that would have been absurd. As noted above, a careful figuring of all the items of value that passed between Holmes and Winans in 1869 (and in part of 1868) points more nearly toward a figure of $1.50.

It should be added that when Winans put in five days at James and Fanny Skinner's house in November, James Skinner soon afterward paid Holmes $10.00. If this was, as seems likely, payment for the work done, it would have allowed $7.50 for Winans and a one-third markup for Holmes. This seems to have been Holmes's practice. In the case of the Hubbard job earlier in the year, Holmes did not indicate what he charged Hubbard for the labor of Herdman and Winans. But the use of the same daily rate for Winans and of a proportionally equivalent $1.00 for Herdman (that is, in either instance the markup is one-third) yields an almost perfect settlement of that difficult account. If the guess is right, as it seems to be, then we may work back to a daily wage of $1.50 for Winans much as we may work back—knowing the answer ahead of time in this case—to Herdman's $.75 daily wage.

Figures on average wages in Westford, which were gathered for the federal and state census in 1850, 1860, and 1865, help to place these calculations in context. In 1850 a day laborer who was not given board was paid an average daily wage of $.88. By 1860 this had risen to $1.00. A carpenter not provided board was paid an average of $1.25 per day in 1850 and $1.50 in 1860. The 1865 figure must be inferred, since the only category for which a wage is cited is farm laborer. The monthly figure (with board) for the summer months was $20.00, said to have increased from an 1860 average of between $15.00 and $18.00. Assuming a similar increase in craftsmen's wages due to wartime inflation, the 1865 wage for a carpenter (without board) ought to have been around $2.00 at the most, and a day laborer should have been earning about $1.30 at the most. No figures exist for Westford in 1870, although the selected retrospective city-by-city figures for carpenters employed in various industries compiled in the 1880 census yield an average for the year 1869 of about $2.40 per day. Common laborers in the same year were earning between $1.00 and $1.75 per day in the same urban-industrial factories.[9] These figures are instructive insofar as they suggest that Holmes was earning just about the average urban carpenter's wages, while Winans was earning the average laborer's. Young Everett Herdman was earning less than what a day laborer had been paid in Westford nearly twenty years before.

It seems reasonable to assume that Judson Winans had worked for or with Holmes in earlier years. Perhaps their earlier association would explain (despite what has been suggested on the question above) why the diary says nothing about Winans's rate of pay. This assumption seems all the more likely when one notes that the diary begins in the midst of things, with Holmes (and Winans) carrying on with the Winans house from 1868. Then Winans just suddenly showed up for work in the Holmes shop in March, without any mention of his having been engaged formally by Holmes. But in the case of Everett Herdman no such carryover seems to have occurred. The new year was promising to be an especially busy one for Holmes. For instance, although it was a common enough practice in the period for workers to labor on Saturdays, in this year (as had not been true for Holmes over the past seven years, when he was serving as Baptist clerk), Holmes was to work the first Saturday of virtually each month, and thus was unable to attend the Baptists' monthly meetings. It may have been this press of business, in fact, that led Holmes to resign from the clerk's position in March. As clerk he was expected to attend every meeting and record what took place, so it is not hard to imagine that Holmes, already busy by March, could predict then that he would have to miss most of the meetings in the coming year. Why not withdraw in advance? Certainly there is no indication in any of his clerk's minutes during previous years that he was disaffected from the church in any way, as, on the contrary, one sometimes notes that Milton Wright was. As it turned out in 1869, Holmes probably attended only the congregation's April meeting.

Everett Herdman's hiring in the month after Holmes resigned as clerk was a response, I think, to the unusually brisk season ahead. But Holmes's use of other workers at various junctures in that season probably represents more nearly the normal practice. Those who put in time on Holmes's own house in 1869 included, besides Holmes, Winans, and Herdman, a large and rather miscellaneous company: the laborer German Baldwin and his son Augustus; Holmes's brother Horace; Holmes's young son Arthur; the farmer and teamster Charles Tyler; the mason William O. Ashley, his "man" (probably the same as his helper "John," although Holmes may have been referring to two different individuals), and Ashley's son, Carlos; the farmers John Goodenough and Edward Rose; Col. Myron J. Hubbard, a farmer, member of the state legislature, and (along with Ashley and Baldwin) one of Holmes's clients later in 1869; Holmes's neighbor and sometime blacksmith, wagonmaker, and mason Denton Salisbury; and—on one occasion, when she "painted [the] chamber"—Mary Shelland Holmes.

Some of these people, or their counterparts, would have helped Holmes on other jobs in a normal year. There is no indication that Holmes undertook any real mason work even on his own house. He picked stone in the quarry and hauled it to the site, hauled brick and sand and lime, on one occasion even mixed mortar for the cellar walls, and may have done some lathing, but at no time did he lay up brick or stone. Nor did he do any plastering in his house or in any other during 1869. Hence, the labor that mason William O. Ashley and his helpers did on the foundation, the chimney, and the interior walls probably would have been hired work regardless of how much pressure Holmes was feeling. Yet it does remain true that Ashley may well have been paying Holmes for some of the debt he anticipated for the house Holmes was to build for him later in 1869. In that case, the exchange of labor here would not have been entirely normal. Holmes did not record paying Ashley, for instance, for work on the Holmes house. He did enter a figure ($126.57) for Ashley's work in the summary of costs (memorandum 6) for his own house. And he included that figure in the total arrived at there, but he probably deducted it from whatever amount he at last charged for the Ashley job. Even less normal was the labor of Myron J. Hubbard, who was not a craftsman himself and must have been helping Holmes to work off part of the bill for his own new house. Or perhaps, since the pay due him for the work was small (Holmes showed it as $15 in the same summary of costs), he was trying to ease Holmes's demanding schedule so that his own job would proceed apace. German Baldwin, on the other hand, might still have performed much of the heavy labor connected with Holmes's new house even if Holmes had not been going to undertake a major job for Baldwin later in the year. But Baldwin's labor for Holmes (credited at $65 in the summary) was also one of the means by which Baldwin would have been enabled to pay for Winans's and Herdman's labor in the fall.

There was a web of exchanged time and effort among these workers, in other words. They may have been linked by a cash economy as well, but swapping was an important economic resource. When Holmes spent the afternoon of May 26 planting corn for Denton Salisbury to pay him back for the 7½ hours Salisbury had spent grading around Holmes's new foundation, he was demonstrating in a small local way just this pervasive pattern.

However these details actually were handled, it is clear that as Holmes entered his heavy building season in 1869 he was prepared to draw on the skills of a sizable number of people in his immediate neighborhood. Although his records for the houses he built for other families

6. William O. Ashley house, 1869, Westford, stonework detail

7. *James C. Holmes house, 1869, Westford, stonework detail*

8. Cemetery Association crypt, 1871, Westford, stonework detail

9. *Commemorative plaque, Cemetery Association crypt, 1871, Westford*

10. Cemetery Association crypt, 1871, Westford

are not as full as that for his own new house, we ought to assume that some of the same people performed in those cases some of the same tasks. Holmes must have served his major clients not as contractor but merely as builder. That is, he must have performed just the woodworking part of any job. Had he been responsible for seeing to the mason work or to the initial excavation, he certainly would have entered details about those tasks in his diary since he would have had to settle with the men who did them. Besides, it is almost certain that in the case of the Ashley job Ashley himself undertook the foundation work and the plastering, much as in the Winans job Winans must have done some of the finish work as well as the rough framing, the roofing, and the siding—working in the latter instances alongside Holmes. Similarly, German Baldwin certainly would have done any heavy labor associated with his job, and it is likely that Hubbard labored on his own job, perhaps plastering and maybe even lending a hand on the carpentry. It is hard to be sure about most of these things, but the pattern of labor, where it can be closely observed, was clearly a mixed one and we probably are right to generalize from that fact.

Then, too, to shift the argument a bit, there is the evidence provided, mute though it may seem, of the foundation of the Ashley house. Exhibiting first-class stone work (fig. 6), it almost certainly bespeaks its owner's special touch. From that point up, the job was mostly an exercise in wood, and that was Holmes's domain. But below the splashboard Ashley expended his own skill as he rarely could afford to elsewhere. For Holmes's house, for instance, he laid a simple journeyman's wall (fig. 7). For Ashley's own house, on the contrary, what he built resembled the fine wall (fig. 8) he almost certainly laid up two years later in the village cemetery. For when Ashley, Hubbard, James's brother John, James's neighbor Denton Salisbury, the produce dealer F. P. Draper, and the wagonmaker and undertaker W. H. Platner (fig. 9), as the officers of the the Westford Cemetery Association, erected a crypt in the graveyard north of the village in 1871, it must have been the group's president, William O. Ashley, who built the small but nicely laid structure (fig. 10). That, too, was a kind of pooled labor fully appropriate to this rural world not yet overcome by a more rationalized modernity.

2 The Holmes Diary for 1869

The diary that James C. Holmes kept during 1869 must have been one of a series in which he recorded his work and his personal life. Efforts to locate others have not proved successful, but this one surviving document does offer insight into how a rural craftsman of the period carried on his labor, as well as how his labor fit into the broader rhythms of his existence.

Much of what I present throughout my book is intended to provide a context for understanding Holmes's apparently modest text. Modesty can conceal a great deal of unarticulated truth that, once extracted, actually proves to be complex. The patterns in Holmes's life were tacit ones in many ways; they did not issue from or issue into public pronouncements but rather came out of and contributed to a vernacular culture that underlay the more obviously textualized public culture of his day. Precisely because the diary is so free of rhetoric and overt ideology, it becomes a rich source for the reconstruction of that vernacular world. As unintentional as a shaving left on the shop floor, it suggests a great deal about the movements implied in its own shape. From this shaving, I have tried to imagine or invent, as it were, the board it came from, and from that board, the tree. At times I have thought it was possible to glimpse the forest where that tree once stood.

The diary should be read not as a literary text or historical document but as a kind of frozen action. Holmes largely recorded what he was going to need to know later, when he settled up with his workers or suppliers or clients. There are some touches that speak to his character, but for the most part the diary is not consciously self-probing or even self-revealing. This is, to my mind, one proof of its veracity: that is, its entanglement with the ordinary activities of Holmes's workaday world lends it a directness it might not have if instead Holmes had embroidered on his life in his prose. I have, in this sense, taken him at his word, seeking to make more explicit in my commentary the implicit patterns he *assumes* in the diary. I have thought of him not as a diarist but as a carpenter who kept a diary. There is a great difference between the two ways of describing him.

What patterns the diary implicates the reader can begin to ferret out in reading it. I think it is important for Holmes's text to be here in its entirety for a number of reasons. How he used language in organizing his labor and his life is part of his story, and I have thought of my own work here as essentially that of a storyteller. A teller must also be a listener, and understanding Holmes—or anyone else—begins with paying attention to him. Although I could have attempted to digest the diary and put what it says merely into my own words, at last such efforts are very problematic. Besides, given my interest in arguing for the full use of documents in the understanding of material culture, I believe that the presentation of Holmes's actual text is a far better proof than any representation of it I might make. Later, in chapter 3 especially, I have my own say about how 1869 must have felt to him. The reader who finds the diary tough going because of its refusal to trumpet itself well might read that chapter first.

A word as to the treatment of the text and my annotations of it is in order here. The diary's text has been altered in a number of ways: (1) all abbreviations except some for measurements have been spelled out; (2) spellings of common words and place names have been corrected except when the spelling may be informative, in which case attention is called to the fact; (3) personal names are spelled exactly as Holmes wrote them (except for the first names of his son Arthur and his helper Everett Herdman, both of which Holmes simply abbreviated; these are spelled out fully whenever they appear); and (4) punctuation, which in the original is composed largely of dashes, has been regularized in accord with modern usage. My notes to the diary seek to explain points raised in the text, especially with regard to Holmes's

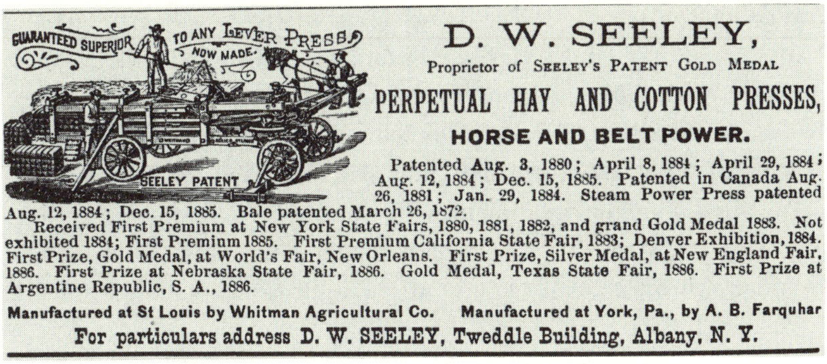

11. David W. Seeley, hay- and cotton-press advertisement,
1883 Albany (New York) Directory

work on the projects of 1869 and his dealings with other individuals. I have identified the people whom he mentions not in the notes but rather in appendix 1, "Westford in 1869: A Biographical Guide," which also includes a discussion of Holmes's family. The reader wishing to know who a particular person mentioned in the text is can find out simply by looking up that person's name.

JANUARY

Friday 1/1. Albany, at Seeley's; hung door and put in window glass. Cold, blustering snow storm. Shoveled snow and worked about the house all day. Gov. Hoffman inaugurated.[1]

Saturday 1/2. Came from Albany to Schenevus, thence home by stage. Paid railroad company 2.05; Stage driver from Schenevus to Westford, 1.20; Mel Griggs, for 5 lbs. Sugar, Box [of] Collars, 1.10. I carry of Mel Griggs, .70.

Sunday 1/3. We attended Methodist Episcopal Church; heard Elder Fox. Text [1 Cor. 7:29]: "But I say brethren the time is short." Rev. Moore called on us.

Monday 1/4. Worked for Winans 3½ hours. Paid Lester Babcock for 1 day with horse, 2.20, also for *Advocate and Journal*, .75.[2] Paid Widow Coats for Cross, 5.00.[3] Received of Clara Holmes on interest due on demand, 50.00.

Tuesday 1/5. Worked for J. Winans 7½ hours on house. Attended library meeting.[4] Paid J. R. Skinner for beef and lumber, 14.00. Warm rain in A.M. P.M. Cold and freezing.

Wednesday 1/6. Worked in shop for self on doors. Filed for John Goodenough and N. Groff.[5] Winans and Wright on house, roof boarding. Paid Geo. Skinner installment.[6]

Thursday 1/7. Worked for J. Winans 7 hours, shingling. I furnish him shingles at 5.00 for M [thousand].[7] I went to Clarksville after Dr. Warren for Clara.[8]

Friday 1/8. Worked for Winans 8 hours on house, shingling. Paid Tyler Brothers for *Tribune*, 1.20;[9] for ½ lb. tea, .63; Mel Griggs, [for] 2 files, .37.

Saturday 1/9. Worked for Winans 4 hours. Filed saws for Wm. Skinner, also for H. W. Holmes. Warm, thawy, rainy day. Worked for Winans 30 hours this week on house.

Sunday 1/10. Attended Congregational Church and Sunday School, heard Rev. Moore. Evening, Uncle Elisha Shelland with us; Arthur [Holmes] and I went to Episcopal Church.

Monday 1/11. Worked in shop for self on doors. Arthur took Mrs. Coats to Roseboom with my horse and cutter. Paid A. Bell for boots, 8.00. Received of E. Shelland, 10.00 on board.

Tuesday 1/12. Worked for J. Winans 6 hours on house; put up stove, hung door. Filed saw for J. Cross.

Wednesday 1/13. Worked for Winans 7 hours. Paid W. H. Platner [for] 143 feet basswood lumber, 2.65. Paid peddler [for] glassware, 1.00. Filed saw for J. B. Skinner. Horace Holmes family visited us.

Thursday 1/14. Worked for Winans 8 hours on house, on piazza. Scaffold fell, wrenched hand, arm, and shoulder. John Marks and family, W. T. Marks and family, and Betsey Hubbard visited us all day.

Friday 1/15. At home with a useless arm. Bought flour and meal at mill. Paid Hall 5.90; S. E. Manzer, [for] medicine and glass, .20; Tyler Brothers, [for] molasses, .47.

Saturday 1/16. Called at Father Holmes's; took Clara up there. Self, wife, and Calvin visited at F. B. Wright's. Worked for Winans 21 hours this week. Filed saw for C. Holmes.

Sunday 1/17. Attended Methodist Episcopal Church; heard Rev. Fox.

Temperance lecture in the evening by Fox in M. E. Church, a large attendage [*sic*] and a good lecture.

Monday 1/18. Worked for Winans 8½ hours in house. Paid Mel Griggs for oil, .25. Fine day but cold.

Tuesday 1/19. Worked for J. Winans 8½ hours, building stairs. Settled with H. E. Draper. Received of Winans for *Examiner and Chronicle*, .75.[10]

Wednesday 1/20. Worked for J. Winans 7½ hours, casing doors. Filed saws for S. E. Manzer and H. Holmes. Cold, blustering day.

Thursday 1/21. Worked for Winans 7½ hours. Received of J. Winans for shingles and lumber, 16.90. Filed saws for N. Roberts and S. E. Manzer. Cash on hand, 33.00.

Friday 1/22. Worked for Winans 7½ hours. Very cold day. Bought of Tyler Brothers 8 lbs. sugar and 11 eggs, 1.50. J. Winans Dr [Debtor] to 39½ hours this week.

Saturday 1/23. Took a load of lumber to Schenevus to get it dressed in Farry's mill. Went to Colliers[ville] by railroad, visited Sash and Blind Factory,[11] thence home. Arthur [was] with me. Brought home load of lumber.

Sunday 1/24. [Attended] Baptist Church, [heard] Rev. H. J. Smith. Text, Luke 24:32. A collection [of] 20.00 for students at Hamilton.[12] Evening, attended preaching at Institute by Moore.[13]

Monday 1/25. A.M., went to Clarksville after Dr. Warren for Will Wright. P.M., Arthur and I cut and drew wood to house. Paid Dr. Warren for medicine, .50; S. E. Manzer, for groceries, 1.07.

Tuesday 1/26. At home filing saws and doing chores. Filed saws [for] F. Campbell, W. E. Wright, Widow Coats, R. S. Hall. Bought shoes for Nettie [Holmes] and goods at Tyler's, 3.00.

Wednesday 1/27. Horace and I went to Colliersville after lumber with two teams. Bought of Barnes & Fox:

1,006	feet flooring	@.025	25.15
533	feet tally[14]	.02	10.56
303	feet inch	.03	9.09
404	feet siding	.035	14.14
70	feet 2-inch	.04	2.80

Thursday 1/28. Worked for J. Winans 8 hours on house, casing win-

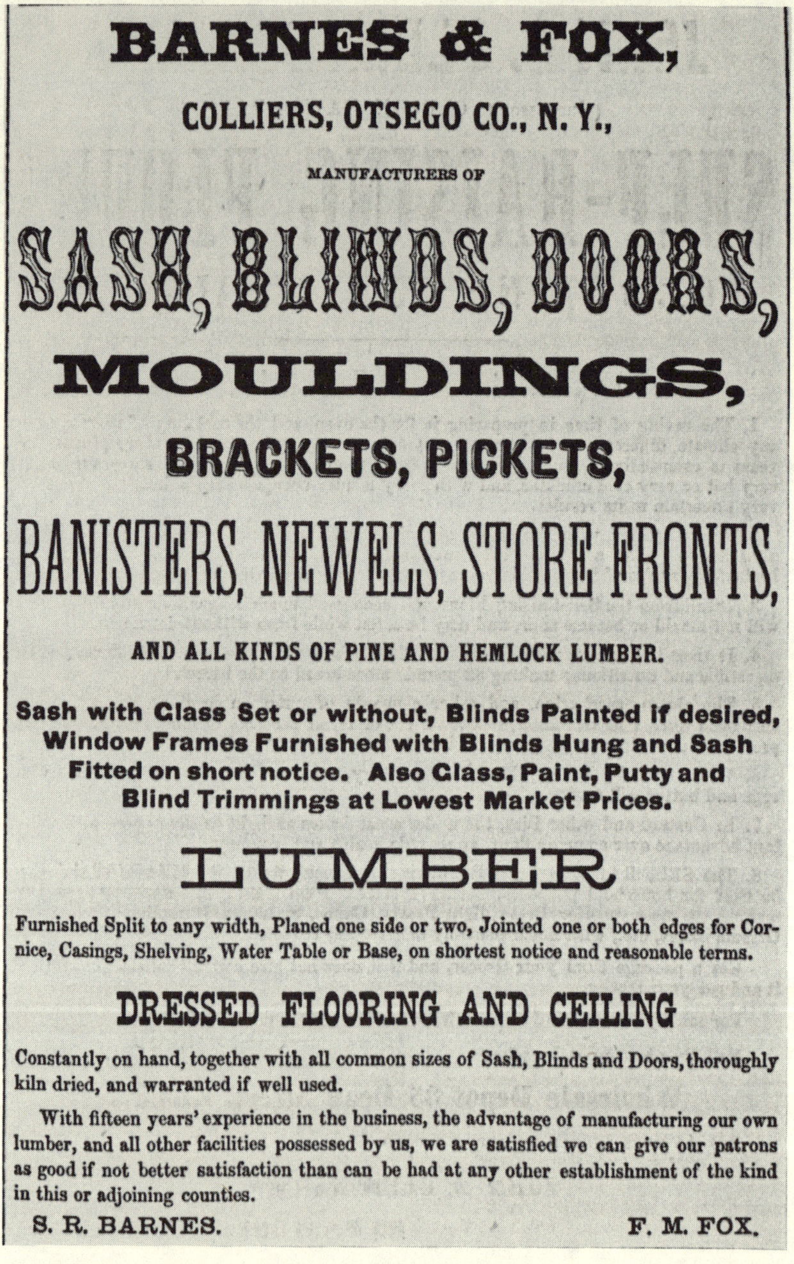

12. Barnes & Fox advertisement, 1868–1869 Albany & Susquehanna Railroad Directory

dows. Filed saw for Gove, .25. Paid J. Cross .40; S. J. Cottrell, [for] use of horse, 1.00.

Friday 1/29. Went to Schenevus, brought home 400 hard brick with one horse.[15] Warm, thawy day. Filed saw for Pank.

Saturday 1/30. Worked for Winans 7½ hours, making 15½ this week. Paid Tyler Brothers for goods, .35. Thawy, rainy day. Bought 100 feet flooring of Winans, .30. Sold him 173 feet [of] hemlock.

Sunday 1/31. At home with children reading Handwriting of God, a description of Jerusalem. Mary heard Rev. Moore in Congregational Church.

FEBRUARY

Monday 2/1. Worked for J. Winans 7½ hours. Cold day. Paid J. Cross .60 on shingles. Sent order to Barnes & Fox for sash, blinds, and doors.[16]

Tuesday 2/2. At home. Filed 2 saws, received .50. Got 2 loads [of] straw of [i.e., from] D. Aldrich. Settled with O. J. Pruyne, 2.96 due me. Cash on hand, 21.80.

Wednesday 2/3. Worked in shop on wardrobe for J. Kelso, 6 hours used. Wm. E. Wright had horse to Cherry Valley after Rev. Lang. Cold and stormy. Received for filing, .50.

Thursday 2/4. Worked in shop for J. Kelso, 6 hours. Filed saws for Ger. Baldwin and Z. Skinner. Daniel Alen died this morning.[17] Received bill of Barnes & Fox.[18]

Friday 2/5. Worked for J. Kelso 4 hours; finished wardrobe. W. E. Wright had horse to Schenevus. Paid Mel Griggs for screws, .53; W. H. Tyler, bolts, .15;[19] W. E. Wright, 5.00, to pay for planing lumber.[20] Sent order No. 2 to Barnes & Fox.

Saturday 2/6. Worked for self, moving shingles and lumber and cutting and drawing wood. Bought ½ lb. tea of S. E. Manzer, .75; ½ gal. oil [of] Tyler Brothers, .25. Republican caucus last evening, Democratic caucus this evening.[21]

Sunday 2/7. Attended Daniel Allen's funeral in [the] Congregational Church. Fine, pleasant day.

13. Congregational church, 1809, Westford

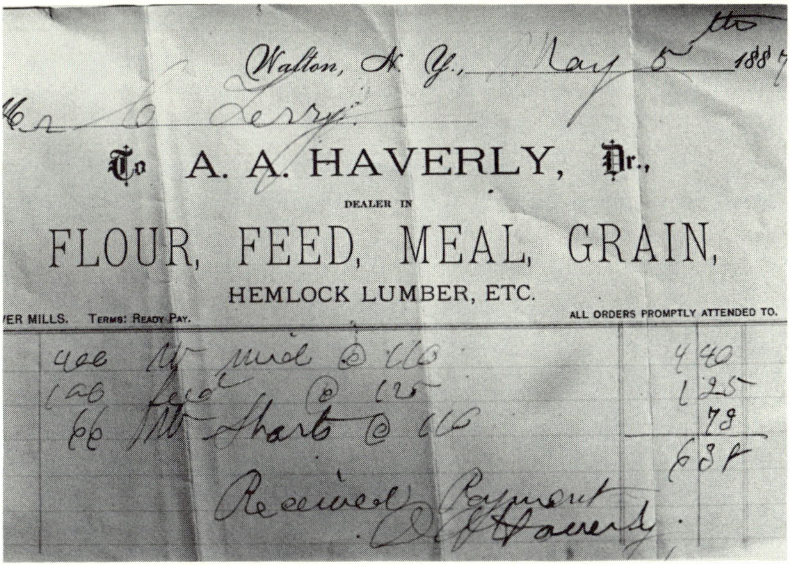
Walton, N. Y., May 5th 1887

To A. A. HAVERLY, Dr.,

DEALER IN

FLOUR, FEED, MEAL, GRAIN,

HEMLOCK LUMBER, ETC.

'ER MILLS. TERMS: READY PAY. ALL ORDERS PROMPTLY ATTENDED TO.

Received Payment

14. Receipt form, A. A. Haverly mill, 1880s, Walton, New York, suggesting local importance of supposedly inferior "Hemlock Lumber"

Monday 2/8. Worked in shop for self, sawing out casing and base for house.[22]

Tuesday 2/9. Town meeting. I worked for self in A.M., putting up shingles. Paid J. Cross 2.00 for making shingles. P.M., attended meeting; filed saw for Murphy.

Wednesday 2/10. Went to Cooperstown with A. I. Griggs as witness to will of Mary Griggs.[23] Paid Cory for sassafras, 2.00. Clara came to town, with us all night.

Thursday 2/11. Tyler estate vendue.[24] Clara and Mary went to C. Holmes's. I went over to stone quarry; German Baldwin and [his] son Augustus at work there for me, commenced yesterday.

Friday 2/12. Worked for Winans 9½ hours in house, casing windows. Chester Wright and wife visited us [for] dinner and supper. Warm, thawy day. Tyler Brothers, butter and sugar, 3.38.

Saturday 2/13. Worked for Winans 9½ hours on house; finished casing doors. Filed saw for W. E. Wright. Warm and thawy. Sold Winans 107 feet hemlock lumber, 1.60.

Sunday 2/14. Attended Baptist Church, heard Elder Ray. Thawy day.

Monday 2/15. Worked for Winans 5½ hours. Filed saws for A. Tucker and J. M. Holmes. Paid J. Cross 2.35 on shingles.

Tuesday 2/16. Worked for Winans 8 hours on house; put up window and door stops, made door sills and put [them] down. Paid Deacon P. Wright for *Examiner*, 2.50. Clara with us.

Wednesday 2/17. Received of A. I. Griggs, 1.30; of Butler, 3.00. Paid C. Holmes 6.00. H. W. Holmes drew 3 loads [of] stone from quarry for me in P.M. Took tea at Col. Hubbard's.

Thursday 2/18. Worked for Winans 7 hours. Called at J. Shelland's. Cold, blustering day.

Friday 2/19. Drew stone in A.M. Horace Holmes drew 4 loads; Charles Tyler, 5 loads. Baldwin and son in quarry.

Saturday 2/20. Cut wood with Will [Wright]. Arthur drew shingles and wood. Piled 1,700 shingles. Received of A. I. Griggs, 1.30.

Sunday 2/21. Attended Baptist Church, heard Elder Ray; also [heard] Vanhoser of Schenevus at Methodist Episcopal Church in morning.

Monday 2/22. Worked for J. Winans 9 hours on house; put up pantry and clothes press shelves.[25] Horace drew 8 loads from quarry. The Baldwins dug stone.

Tuesday 2/23. Worked for Winans 8 hours on columns and cellar stairs. Cold and windy. Arthur got 2½ lbs. butter of J. B. Skinner. Sold J. Winans 46 feet [of] boards [@] .035, 1.61.

Wednesday 2/24. Worked for Winans 7 hours on banisters and handrail. Very cold and blustering.

Thursday 2/25. Went to Schenevus, got blinds and doors, bought ½ barrel flour of Thompson, 4.63. Bought of Tyler Brothers, molasses and eggs, .93. Filed saw for Gove, .25.

Friday 2/26. Received of W. H. Tyler, 15.00. J. B. Skinner: paid him for 2½ lbs. butter, 1.00. Received for filing: Patrick, .30; [Stephen] Bentley, .25. Filed saw for Z. Skinner. Called on Father Holmes.

Saturday 2/27. Went to Colliersville with Hubbard and Ashley via railroad from Schenevus. Ticket, .30; bill at hotel, .75. Paid Chase & Mills .85. Paid J. Cross on shingles, 5.25. Went at Hubbard's request.[26]

Sunday 2/28. Attended Methodist Episcopal Church, heard Rev. Fox. Father Holmes sick; bilious fever.[27]

MARCH

Monday 3/1. Called on Father Holmes, who is sick. P.M., Arthur and I drew stone out of quarry. Paid C. Tyler for drawing 8 loads, 2.25; Sherm Manzer, 6.00.

Tuesday 3/2. Attended church; heard Bishop Duan, D.D., in the Episcopal Church. Charles Tyler drew 10 loads stone for me from quarry; paid him 5.00. Set up with Father all night.

Wednesday 3/3. Took Father and Mother Sheldon to Charles Davis's. Arthur and I cut and drew wood to house. Bought ⅜ cord hard maple, seasoned wood, of Geo. Wing, 2.00 on account.

Thursday 3/4. Worked in shop for self on window frames 10 hours. Evening, settled with Daniel Aldrich; received of him, 38.00.[28] Winans had horse to Bentley Hollow.

Friday 3/5. Worked for self in shop on window frames. Received of J. Kelso, 5.00 on work.[29] Evening, at home.

Saturday 3/6. Worked for self in shop on window frames. Lent W. E. Wright 15.00 till April 1st.[30] Set up with Father Holmes all night.

Sunday 3/7. At Father Holmes's in A.M. Washed and dressed him. P.M., at home.

Monday 3/8. Worked for self in shop on window jambs. Sat on jury 2 hours on D. Drake's case of insanity.[31] Paid M. Griggs 1.45 for groceries.

Tuesday 3/9. Went to Cooperstown for A. I. Griggs to know will of his mother.[32] Stopped at Kise's Hotel. Came home with Wm. Marks.

Wednesday 3/10. Moved into the north part of the house.[33] Worked in shop for self 8 hours. Rainy day.

Thursday 3/11. Worked in shop. D. W. Seeley and wife came to town, stopped with us overnight. Arthur drew 5 loads manure on the farm.[34]

Friday 3/12. Worked in shop on frames for self. Arthur took Seeley's folks up to J. B. Skinner's.

Saturday 3/13. Went to Schenevus. Took Seeley and wife to depot. Bought boring machine of Chase & Mills, 9.50.[35] Set up with Father Holmes all night. He is gaining.

Sunday 3/14. At home with children. Arthur took Mary to Father's and brought Clara back. Will and Sarah went home with J. B. Skinner.

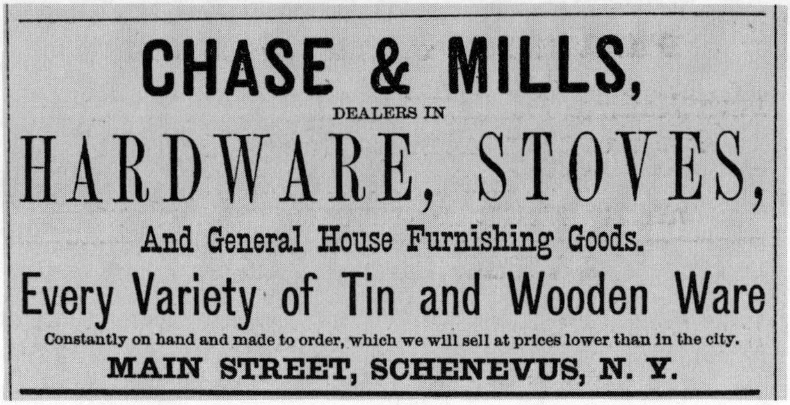

15. *Chase & Mills advertisement,* 1868–1869 Albany & Susquehanna Railroad Directory

Monday 3/15. Worked in shop on sash base and casings. Received of J. M. Holmes on settlement, 8.00.[36]

Tuesday 3/16. Bought 25 acres' land back of [i.e., from] D. Aldrich; received of him, 40.00.[37] Worked in shop ½ day. Winans worked in shop 8½ hours.[38] Called at Father's in evening; he is worse.

Wednesday 3/17. Worked in shop ½ day. Called at Father's, who is much worse. Paid S. E. Manzer .13; Tyler Brothers, .15. Cream cow had a calf.

Thursday 3/18. In shop ½ day, filing up boring machine. Called on Father, who is some better but is very feeble. Paid H. W. Holmes .31.

Friday 3/19. Called up at 4 o'clock by Marvin to go up to Father's, who is just alive. Went up soon as I could; stayed all day and all night. Father cannot live.

Saturday 3/20. Sent telegraph [*sic*] to A. B. Chapman. Father is dying. He passed away at 1 o'clock P.M. Gone to his reward in heaven.

Sunday 3/21. We attended Methodist Episcopal Church; heard Rev. Ray of Waterville. Called at Horace's. Received answer to telegram: they cannot come.

Monday 3/22. Attended Father's funeral. Sermon by Rev. Fox, text: "Let me die the death of the righteous and let my end be like His." A full house.

Tuesday 3/23. Received of M. Agans, [for] filing, .45; of Pank, .30. Filed saw for C. Wright, .30. Settled with W. E. Wright.[39] Received letter from A. B. Chapman. Winans worked in shop.

Wednesday 3/24. Worked in shop all day; Winans also. Received of McClintoc, .30 for filing. Gave Mary 1.50 to buy dress. Sent letter to Mrs. Fanny Roberts, Ill[inois].

Thursday 3/25. Took Anna Holmes to Horace's. Worked in shop the rest of the day. Arthur [was] sick; I had the chores to do.

Friday 3/26. Worked in shop 8 hours; Winans worked 5 hours. Rainy all day. M. J. Holmes and Clara Holmes called on us.

Saturday 3/27. Worked in shop 5 hours; Winans, 9 hours. P.M., took Anna Holmes over to Parker's [in] Decatur. Thawing all day. Received of Tyler Brothers, 25.00. Paid C. Tyler 1.00.

Sunday 3/28. Attended church, heard Rev. Fox. Text: "I was glad when they said unto me, 'Let us go into the house of the Lord.' " Mary and I called at C. Davises'.

Monday 3/29. Worked in shop for self on cornice and casings. Bought 24 lbs. ham of Horace [Holmes]. W. E. Wright had horse to Schenevus.[40] Rainy all day. Father Shelland died.

Tuesday 3/30. Bought cow of Horace Holmes, 75.00. Bought shroud of W. H. Platner and laid out D. Shelland.[41] Stormy all day.

Wednesday 3/31. Attended Father Sheldon's funeral at Worcester. Wife [was] sick and could not go. I took Mother and Orilla. Very bad going.[42]

APRIL

Thursday 4/1. Received of Tyler Brothers, 53.50. Paid and settled [with] Tyler Brothers, 3.90.[43] Paid Dolly Bentley interest, 28.00. Paid R. S. Hall's saw bill, 35.00.[44] Paid R. S. Hall [for] 500 lbs. bran, 8.00. Received of Geo. Wing, 5.00. Moved hay and wood.[45]

Friday 4/2. Filed two saws, [for] S. Manzer and A. Tucker. Received of H. W. Holmes, 150.00. Lent Winans 40.00.[46] Working, moving lumber. In shop 3 hours. Gave Horace Holmes note [for] 150.00, due in 1 year.[47]

Saturday 4/3. Worked for self in A.M. P.M., self and wife attended Church meeting; 22 present. Elected O. J. Pruyne deacon for 2 years.[48]

Sunday 4/4. We all attended Church; heard Elder Fox preach his last sermon.[49] Text: "What I have written I have written."

Monday 4/5. Worked in shop; made ladder 19 feet long.[50] Sold calf to C. Tyler, 2.50.

Tuesday 4/6. Worked in shop. Filed saw for R. S. Hall, .40. Snowy and thawy. In shop 3 hours.

Wednesday 4/7. Drew manure part of the day; Arthur helped. Worked in shop 4 hours. Bought of H. W. Holmes 6 bushels oats. Received of R. S. Hall, .50.

Thursday 4/8. Worked in shop 8 hours. Arthur drew 6 loads manure. Paid W. H. Tyler 1.15. I filed saw for H. Groff. Received of C. Tyler, 2.50. Received of J. Winans, 40.00 lent [on April 2].

Friday 4/9. Arthur and I sawed wood in A.M.; drew manure in P.M. Sent draft [for] 50.00 to Barnes & Fox; paid the money to W. H. Tyler.[51]

Saturday 4/10. Filed saw for self. Worked for M. Griggs 2 hours, hanging blinds; put on door bell. P.M., painted 12 window sash for self.[52] [Bought] 4 lbs. paint of M. Waterman.

Sunday 4/11. Attended Congregational Church; heard Rev. Moore.

Monday 4/12. Bought cow of Horace Holmes and brought her home, 75.00. W. E. Wright had horse to Schenevus. Filed saw for F. Campbell, .30. Arthur and I cut wood in P.M.

Tuesday 4/13. Everett Herdman commenced work for me at 18.00 per month. Moved 2 loads of hay to Susan Chester's barn.[53] Drawed 2 loads wood and 1 of manure. P.M., Everett and I sawed wood.

Wednesday 4/14. Sawed wood and hewed timber all day.[54] Everett helped. Arthur drew 6 loads [of] manure. Paid W. H. Tyler [for] fish, etc., .59. Fine day.

Thursday 4/15. Worked on timber. Everett 1 day with me. Arthur drew 6 loads manure. Bought 4 lbs. sugar @ .17, .68. Fine day.

Friday 4/16. Everett and I worked on timber sills [for] framing house. Fine day.

Saturday 4/17. Worked on timber in barn and shop. Everett all day, 5 days this week.[55] Paid W. H. Tyler 1.43; S. E. Manzer, .87.

Sunday 4/18. Attended Congregational Church; heard Rev. Moore.

Monday 4/19. Worked on frame. Everett with me 1 day. Walter Coats and Arthur drew manure. Paid E. R. Treat .60; Mary Holmes, .60.

Tuesday 4/20. Everett and I worked on frame all day. Walter and Arthur spread manure and drew lumber. Bought lumber of Hall:

Bought 16 sticks 2 inches × 7 inches × 20 feet
Bought 4 sticks 2 inches × 7 inches × 16 feet
Bought 34 sticks 2 inches × 4 inches × 12 feet
Bought 127 feet [of] boards

Wednesday 4/21. Worked in shop in A.M. Everett ½ day. Rainy. Called at W. T. Marks's. Evening at home.

Thursday 4/22. Went to Schenevus in A.M. P.M., worked on timber. Everett 1 day. Bought of Chase & Mills 1½ dozen panes [of glass], etc.; paid 7.75. [Bought something] of Sarah Wright; paid 1.00.[56] Brought home sash.

Friday 4/23. Worked on house frame and moved timber all day. Everett 1 day for self.

Saturday 4/24. Worked on frame in A.M. P.M., fitted and hung 5 sets blinds for Lester Babcock. Everett worked all day. Paid S. E. Manzer .63; D. Salsbury, school tax, .30.[57]

Sunday 4/25. At home. Mary and children attended Methodist Episcopal Church; heard Rev. Fox.

Monday 4/26. Everett and I worked on shop frame 1 day.[58] Pleasant but windy. Received of Wm. Skinner, .70. Paid W. T. Marks, .20; S. E. Manzer, .76. Milton Draper [and] folks to tea.

Tuesday 4/27. Worked on shop frame and moved timber. Everett helped 1 day. Evening, severe pain in left knee; restless all night.

Wednesday 4/28. Worked in shop in A.M. P.M., worked on frame. Everett 1 day.

Thursday 4/29. Worked on shop frame; Everett helped. Bought sack flour of R. S. Hall, 2.50.

Friday 4/30. Commenced digging cellar. German Baldwin with team. We plowed and scraped all day.

MAY

Saturday 5/1. Worked in cellar in A.M.; snowy. P.M., Everett and I worked in shop on windows. Bought 4 lbs. paint of M. Waterman, .75.

Sunday 5/2. Attended Methodist Episcopal Church; heard Rev. Fox. Snow storm all day.

Monday 5/3. Bought ⅓ ton hay of N. Bidlake at 12.00 per ton. Worked for M. Griggs: self and Everett 3 hours each in P.M. Snow storm all day.

Tuesday 5/4. Worked in shop in A.M. P.M., in cellar digging and scraping. Ground covered with snow. Bought 2 boxes glass, 9 × 12, of M. Griggs. Bought of W. H. Tyler, coffee and rice.

Wednesday 5/5. Worked in cellar digging. German Baldwin helped 1 day. Paid M. Griggs for glass, 8.75; W. H. Tyler, sugar, .52.

Thursday 5/6. All hand [*sic*] in the cellar. G. Baldwin and John Goodenough helped. I hung 3 sets blinds for L. Babcock.

Friday 5/7. Finished digging cellar. Everett and Baldwin worked all day. Received 4 window sash of Barnes & Fox.

Saturday 5/8. W. T. Marks's team and Free[?] plowed for me one day. Baldwin worked ½ day, Everett 1 day. Paid J. Cross 1.50 on shingles.

Sunday 5/9. Attended Baptist Church; heard Perry Bentley preach for the first time. An excellent sermon well delivered.

Monday 5/10. Ashley and man began mason work on cellar.[59] Arthur and I drew stone. J. Winans and Everett on shop. Sold cow to Lucy [Holmes Marks], received of her 70.00 and [a] calf.

Tuesday 5/11. Ashley and hand on cellar. Winans and hand [Everett Herdman] on shop. Arthur and I drew stone. Planted ½ bushel potatoes. Fine day. Bought 18½ lbs. nails of J. Winans.

Wednesday 5/12. Ashley and John worked on cellar. Judson and Everett on shop. Arthur and I drew stone. Paid L. Babcock for lumber, 10.00.

Thursday 5/13. Drew stone [and] shingles and planted potatoes. Masons off. Rainy P.M. and all night.

Friday 5/14. Drew stones for cellar. Ashley and man worked. Paid J. Cross on shingles, 2.00.

Saturday 5/15. A.M., worked in shop; hung door for Mrs. Coats and

16. Melville Griggs's store, ca. 1830, Westford

A. Chester.[60] P.M., went to Schenevus after lime. Came home in the rain. Paid M. Wright for posts, 2.50.[61]

Sunday 5/16. Attended Methodist Episcopal Church. Heard Rev. P[omeroy] Wright.

Monday 5/17. Everett and I worked on shop. Leander Roberts came to town after [his] mother. Ashley worked ½ day.

Tuesday 5/18. Worked on shop, finished it; ready to move into. Rainy.

Wednesday 5/19. Moved in A.M. P.M., picked and drew stone for cellar. Paid L. Roberts 120.00.[62] German Baldwin helped with horse. Ashley 1 day.

Thursday 5/20. Everett and I mended fence in A.M. P.M., Everett attending Ashley; I drew sand, Rose drew 1 load. Ashley worked 1 day. Received of L. Snyder, 100.00.[63]

Friday 5/21. Worked on cellar. I drew 2 loads sand, Rose drew 2 loads, in A.M. P.M., Planted potatoes. Ashley and Everett 1 day on cellar. Paid Everett 5.00.

Saturday 5/22. Everett attended Ashley on cellar 1 day. Built backhouse.[64] Furrowed out corn ground. Paid J. Kelso 20.00; W. T. Marks, 100.00.[65]

Sunday 5/23. Attended Episcopal Church; heard Rev. Hobs. Also, Methodist Episcopal Church in evening; heard Fox lecture on evil speaking.

Monday 5/24. Ashley worked on cellar 1 day. Everett on the same, and set fence posts. Paid O. Tyler for 15 bushels lime and drawing, 7.50; Wm. Utter, for brick, 1.00. Received [of] D. Kelso, .50.

Tuesday 5/25. Ashley and Everett on cellar. I mended fence and built pigpen.

Wednesday 5/26. Ashley and son Carlos on cellar 1 day. I planted corn for D. Salisbury in P.M. to pay him for ½ day on cellar. Paid E. R. Treet [for] meat, 1.50.

Thursday 5/27. Ashley and Carl finished on cellar. Mary went up to Marvin's and Markses'. I mended fence. Bought 2 pigs of Horace, 6.50.

Friday 5/28. D. Salisbury worked ¾ day grading about cellar. We all helped. Put on sills. Rained about 3 o'clock. Paid N. Barnard 1.00; W. H. Tyler, [for] nails, .44.

Saturday 5/29. Raised house; D. Salisbury, J. Winans, Everett helped 1 day each. Drove off with the rain. Paid W. H. Tyler 4.12.

Sunday 5/30. Attended Methodist Episcopal Church; heard Rev. Fox. Went to Congregational Sunday School.

Monday 5/31. Worked on house. Everett and Winans helped. Ashley had horse and wagon to Cooperstown. Sent by Ashley, 3.80.[66]

JUNE

Tuesday 6/1. Worked on house and in shop; Jud and Everett helped. Rainy. Gave Mary H[olmes] 1.50.

Wednesday 6/2. Worked on house; Judson and Everett put on siding. Paid John Goodenough 2.00 for 1 day's digging. Bought of W. H. Tyler, 10 lbs. nails, .60.

Thursday 6/3. Worked on house, putting on siding. Judson and Everett worked 1 day each. Fine day.

Friday 6/4. Judson, Everett, and self on house 1 day.

Saturday 6/5. Worked in shop. Everett 1 day. Judson ½ day. Paid F. P. Draper 6.00.

Sunday 6/6. Attended Methodist Episcopal Church; heard Rev. Daniels. Visited the cemetery. Paid H. W. Holmes 5.00.[67]

Monday 6/7. Worked on house. Jud, Everett, and self finished siding. Paid R. S. Hall [for] flour, 8.75; W. H. Tyler, nails, .78.

Tuesday 6/8. Worked on house, roof boarding and shingling. Jud and Everett 1 day each. Bought 50 lbs. shingle nails of M. Griggs.

Wednesday 6/9. Worked on house. Everett and Winans helped 1 day each.

Thursday 6/10. Worked on house, shingling. Everett and Jud 1 day each. Made cellar window frame[s] for M. J. Hubbard, 3 hours. Got bunch of shingles of C. Holmes estate.

Friday 6/11. Finished shingling. Everett and Jud 1 day each. Paid J. Cross for shingles, 3.00. Bought shingles of H. Bell, 1.50.

Saturday 6/12. Worked on house, laying floor. Winans painted in P.M. Winans and Everett 1 day each. Bought of W. H. Tyler, 50 lbs. paint, 2 gal. oil, 10.00. Received of W. H. Tyler, 25.00.

Sunday 6/13. At Methodist Episcopal Church; Rev. Fox.

Monday 6/14. In house laying floor. Everett and Jud 1 day. I made 4 window frames for M. J. Hubbard, 4 hours. Bought 50 lbs. white lead and 18 glass of W. H. Tyler; 200 brick [of] W. T. Marks.

Tuesday 6/15. We all worked on house. Judson painted outside. Paid R. S. Hall for 5 days' road tax, 5.00; S. E. Manzer, fish, .53, alcohol, .25.

Wednesday 6/16. Worked on house. Judson finished painting outside. Everett worked 1 day. 50 lbs. lead, 3 gal. oil for first coat.

Thursday 6/17. Everett and I worked in house; moved in bench and tools. I put up stairs. I divided line fence with Wm. O. Wilbur; he makes upper end.[68]

Friday 6/18. Worked on house. Everett ½ day. Arthur and I mixed mortar in P.M. for cellar wall.

Saturday 6/19. Worked on house, putting up partitions. Everett 1 day. Filed saws for McClintock, N. Roberts, and W. T. Marks. Paid S. E. Manzer 1.88.

Sunday 6/20. At home.

Monday 6/21. Everett, Jud, and self on house all day. Carlos Ashley plastered cellar. Paid W. H. Platner 2.00; H. Bell, for shingles, 1.50; S. E. Manzer, .23.

Tuesday 6/22. Everett, Jud, and self in house all day. Paid H. W. Holmes 1.50, balance due on pigs.

Wednesday 6/23. Jud, Everett, and self on house; base and casings. Received of W. H. Tyler, 10.00. Paid Everett 2.00. Finished chamber.

Thursday 6/24. Judson, Everett, and self in house 1 day. Fine day. Mary painted chamber. Bought of Horace Holmes, 2 bushels potatoes, 15¾ lbs. pork.

Friday 6/25. Everett, Jud, and self on house. Everett and I hoed potatoes and corn in P.M. Bought ¼ bushel swordfish of S. E. Manzer, 4.00.

Saturday 6/26. Everett, Winans, and self worked in house 1 day. 6 days each this week. Carlos and John lathed in A.M. Rainy morn. Paid W. H. Tyler .75.

Sunday 6/27. At home. Rainy all day.

Monday 6/28. Self and hands on house 1 day. John and Carlos lathed. Bought 25 lbs. nails of W. H. Tyler, 1,000 lath of R. S. Hall.

Tuesday 6/29. I worked for Ashley on cellar window frames. Jud and Everett for me on house 1 day each. 375 lath from Hall's.

Wednesday 6/30. Finished up on my house. Jud and Everett worked; Jud lathed, Everett dressed fence rails.

JULY

Thursday 7/1. Everett, Jud, and self worked on M. J. Hubbard's frame 1 day each. Sent for glass and [putty] tubes by Horace; bill 4.75. Received of W. H. Tyler, 15.00. Paid S. E. Manzer 1.15.

Friday 7/2. Self, Everett, and Judson worked for Hubbard 1 day each on frame. Sent letter to D. Chester. Leroy Marks and family with us. Paid S. E. Manzer .40.

Saturday 7/3. Worked for M. J. Hubbard. Jud, Everett, and self 1 day each. Paid Everett 3.00.

17. Baptist church, 1836, Westford

Sunday 7/4. Attended Methodist Episcopal Church; heard Rev. Fox, and Cob, presiding elder, preached in evening.

Monday 7/5. All hands off. Winans and I went to Schenevus. Paid J. Winans 5.00; W. E. Wright, 5.00; Chase & Mills, .90.

Tuesday 7/6. Worked for Hubbard. Self, Winans, and Everett 1 day each on frame.

Wednesday 7/7. Self and Everett 1 day each. Winans off [for] Association in Clarksville.[69]

Thursday 7/8. All hands for Hubbard raising. John and Carlos ¼ day lathing for me. Rainy P.M.

Friday 7/9. Worked for Hubbard. Self, Winans, and Everett 1 day each. Carlos and John lathed ¾ day each.

Saturday 7/10. All hands for Hubbard; finished raising frame. Paid M. Griggs .60.

18. Draper's Hall, built 1823 as Methodist church, photo ca. 1860, Westford

Sunday 7/11. Attended Baptist Church; heard a stranger try to preach.[70]

Monday 7/12. Self, Everett, and Winans on Hubbard's house 1 day each. Fine day, no rain.[71]

Tuesday 7/13. Self, Jud, and Everett 1 day for Hubbard. Rainy P.M.

Wednesday 7/14. Self, Jud, 1 day each. Everett ½ day for Hubbard. Sabbath School picknick–knack [*sic*] at Draper's Hall.[72]

Thursday 7/15. Worked for Hubbard; self, Jud, and Everett one day each. D. Cumming had horse to Middlefield over night.

Friday 7/16. At Hubbard's, all hands 1 day each; put up cornice to wing and shingl[ed]. Paid M. Griggs 8.00; Rose, for lime, 15.00; received of Tyler, 15.00.

Saturday 7/17. All hands at Hubbard's, 1 day each.

Sunday 7/18. Attended Baptist Church; heard Rev. J. Parsons. Evening, self and wife went to Bentley Hollow;[73] heard Rev. Perry Bentley.

Monday 7/19. All hands on Hubbard house this day. Fine hay day; bought scythes of Tyler, 1.75. Sent letter to Clara.

Tuesday 7/20. All hands at Hubbard's. Paid E. R. Treet 2.63; J. Tipple, use of barn, 1.00.[74]

19. Methodist Episcopal church, 1863, photo ca. 1940, Westford

Wednesday 7/21. Self and Everett at Hubbard's, ½ day each; Winans, 1 day. Everett and I at house in P.M., fixing barn for hay. Paid Wing for hay, 1.00.

Thursday 7/22. All hands at Hubbard house all day. Aldrich does my haying, for [a] ⅖ [share].

Friday 7/23. All hands at Hubbard's. Aldrich drawing hay from back meadow. John and Louisa visited us.

Saturday 7/24. All hands at Hubbard's. Bought bill of goods at Tyler's, 2 lbs. nails, ½ lb. tea, soap.

Sunday 7/25. Attended Congregational and Episcopal Church; heard Dr. Hobs of New Jersey, Rev. Swinerson of Cherry Valley.

Monday 7/26. Everett and Winans worked for me in hay field [in] A.M.; P.M., on Hubbard house ½ day each. Hubbard, Ashley, and Carlos worked for me 1 day each on house; mortar[ed] chimney.

Tuesday 7/27. Self, Winans, and Everett on Hubbard house 1 day each. Paid J. Tipple 1.00.

Wednesday 7/28. All hands at Hubbard's. Hubbard and Carlos mixed mortar for me ¾ day. Ashley finished brickwork for me. C. Wright and wife with us to tea.

Thursday 7/29. All hands at Hubbard's; finished shingling, moved bench and tools [inside]. C. Wright and wife had horse and boy [i.e., Arthur Holmes] to [visit] H. Peobles and Jno., [in] S[chenevus]. Rainy last night and this morning. Methodist Episcopal Church [installed] new bell.[75]

Friday 7/30. Winans and self 1 day each, Everett ½ day, for Hubbard. Everett helped Aldrich in P.M. get in 4 loads hay. Aldrich had my horse.

Saturday 7/31. All hands for Hubbard in A.M. P.M., off getting hay; sold Winans 2 tons. Paid M. Griggs's bill, 1.00; S. E. Manzer, 2.00. Everett for Aldrich ½ day.

AUGUST

Sunday 8/1. Attended Methodist Episcopal Church and Baptist Church; Rev. Powers and young Lowell. Evening at home.

Monday 8/2. Self and Winans 1 day on Hubbard's house; Everett ½ day. Ashley and son plastered for me; Hubbard attended ½ day.

Tuesday 8/3. Worked on Hubbard's house; self, Winans, and Everett 1 day each. Ashley, Hubbard, and Carl on my house, plastering.

Wednesday 8/4. Winans and Everett 1 day each, self ¾ on Hubbard's. Ashley, Carl, and Col. [Hubbard] on my house. Rainy P.M. and evening.

Thursday 8/5. Winans and self on Hubbard's house 1 day each. Everett helped make mortar; Ashley, Carl, and Hubbard plastering my house. Paid R. Hall 2.70.

Friday 8/6. All hands in Hubbard's house casing up, 1 day each. Ashley, Carl, and Hubbard on my house.

Saturday 8/7. Self and Winans for Hubbard 1 day each. Hubbard, Ashley, and Carl for me ½ day each; finished plastering. Received of Hubbard, 10.00.

Sunday 8/8. We all attended Church; heard ———.[76] Methodist Episcopal Church in evening; H. Fox.

Monday 8/9. All hands on Hubbard's house 1 day each. Paid R. S. Hall, 1 barrel flour, 9.00.

Tuesday 8/10. All hands on Hubbard house 1 day each. Fine day. Paid E. R. Treet 1.80.

Wednesday 8/11. All hands at Hubbard's 1 day each. Evening, at prayer meeting at D. Aldrich's.

Thursday 8/12. All hands at Hubbard's 1 day each. Fine day.

Friday 8/13. All hands at Hubbard's 1 day each. Sold Hubbard 218 feet partition lumber.

Saturday 8/14. All hands at Hubbard's. Received of Hubbard, 15.00. Hubbard's wife and boy had horse and wagon to Schenevus all day. Bought 3 pictures, 3.00.[77]

Sunday 8/15. Sister Clara with us. She and Mary attended Church in A.M. I was at home with children. I attended Congregational Church in evening; heard Rev. Moure.

Monday 8/16. All hands at Hubbard's 1 day each. Fine cool day.

Tuesday 8/17. All hands at Hubbard's 1 day each. Rainy. Paid L. Manser for 11 bushels lime and 10 bushels sand.

Wednesday 8/18. All hands at Hubbard's. Everett off in P.M. Sold 103 lbs. [of] butter, 30.90.[78] Paid M. Griggs 17.40; W. H. Tyler, 8.10; S. E. Manzer, 1.30; Everett Herdman, 1.00.

Thursday 8/19. All hands at Hubbard's 1 day each. Very warm: first hot day this simmer [*sic*].[79] Paid H. W. Holmes 4.75 for glass and [putty] tubes.

Friday 8/20. All hands at Hubbard's 1 day each. Everett off 4 hours in P.M.; had horse to Worcester.

Saturday 8/21. All hands at Hubbard's 1 day each; finished for the present. Winans 42 days, Herdman 40 days, self 44 days [first written 43½].

Sunday 8/22. Attended Baptist Church; heard Elder Rowland. Prayer meeting at 4 o'clock and 7½. Orilla and Mother Shelland with us.

Monday 8/23. Commenced Wm. O. Ashley's house. All hands 1 day each on frame. Paid M. Griggs .88; S. E. Manzer, 1.56.

Tuesday 8/24. All hands at Ashley's on frame 1 day each. Paid S. E. Manzer for sand, 3.00.

Wednesday 8/25. All hands at Ashley's 1 day each. Rained at 4 o'clock. Received of J. Thompson, .70 for use of horse.

Thursday 8/26. All hands at Ashley's 1 day each on frame; put on sills. Paid H. W. Holmes 1.00 for mutton.

Friday 8/27. All hands at Ashley's 1 day each; put up frame.

Saturday 8/28. All hands at Ashley's 1 day each. Received of M. J. H[ubbard], 25.00. Paid J. Winans 10.00; E. Herdman, 7.00; S. E. Manzer, .63; Menzo Tyler, 5.00.

Sunday 8/29. We all attended Baptist Church; heard Rev. Rowland preach. Prayer meeting at 4 P.M. and evening.

Monday 8/30. All hands at Ashley's 1 day each. Winans put on water table. Mary at Hull's picking hops. Arthur sick.

Tuesday 8/31. All hands at Ashley's on house 1 day each.

SEPTEMBER

Wednesday 9/1. All hands at Ashley's 1 day each. Bought firkins of J. Thompson for H. Holmes.[80]

Thursday 9/2. All hands at Ashley's 1 day each. Sold H. Holmes firkins, 1.50.

Friday 9/3. All hands at Ashley's. M. J. Hubbard had horse and wagon by Charley. Paid J. Winans 5.00.

Saturday 9/4. Worked at home for self. Everett ½ day. Winans off to Colliersville.

Sunday 9/5. Attended Baptist Church; heard Elder Ray of Waterville. Communion.

20. Hop picking in Otsego County, 1905

Monday 9/6. Worked for Ashley 1 day. Winans and Everett at [German] Baldwin's 1 day each.

Tuesday 9/7. At Ashley's 1 day. Everett and Jud at Baldwin's. G. Baldwin had 2 sticks of timber of me, 12 × 24. Everett had horse to Worcester.

Wednesday 9/8. At Ashley's 1 day. Everett and Jud at Baldwin's. Daniel Greene helping me on Ashley's job.

Thursday 9/9. At Ashley's 1 day. Everett and Jud at Baldwin's 1 day each.

Friday 9/10. At Ashley's 1 day. Everett and Jud at Baldwin's 1 day each. J. Thompson had horse to Cooperstown.

Saturday 9/11. At Ashley's 1 day. Everett and Jud at Baldwin's. Received of J. B. Skinner, 100.00; received of M. J. Hubbard, 50.00. Paid Ev Herdman 10.00; paid J. Winans 20.00. Gave J. B. Skinner [a] note [for] 100.00, due first [of] April next.

Sunday 9/12. Attended Baptist Church; heard Rev. H. Rowland three times—once in the Hollow and twice in our church. Evening, a full house; [he preached on] the 2 covenants, a glorious discourse.

Monday 9/13. At Ashley's 1 day on house. Everett and Jud at Baldwin's.

Paid W. H. Tyler 6.00 for boots. L. F. Preston mended shoes, 1.25.

Tuesday 9/14. At Ashley's 1 day. Everett and Jud at Baldwin's 1 day each.

Wednesday 9/15. Went to Cooperstown [for] County Fair. Bought draft at bank, 6%.[81] Expenses in town, .70.

Thursday 9/16. At Ashley's ½ day. Winans at Baldwin's. Aldrich got in [my] oats.

Friday 9/17. At Ashley's 1 day. Winans and Everett at Baldwin's 1 day each.

Saturday 9/18. At Ashley's 1 day. Winans and Everett at Baldwin's 1 day each. Evening, met at [S. J.] Cotrell's to sing.

Sunday 9/19. At Baptist Church; heard Elder Rowland. Evening, at Congregational Church; heard Dr. Campbell of New York.

Monday 9/20. At Ashley's 1 day. Everett and Winans at G. Baldwin's 1 day each. John Mason died.[82]

Tuesday 9/21. At Ashley's 1 day. Winans and Everett at Baldwin's 1 day. J. Mason's funeral. Paid E. R. Treet, beef, 1.20.

Wednesday 9/22. At home; worked about the house. Winans and Everett at Baldwin's 1 day each. Rainy all day. Paid M. Griggs 6.00.

Thursday 9/23. At Ashley's 1 day on house as usual. Winans and Everett at Baldwin's. Received of J. Snyder, 1.25. Paid S. E. Manzer 1.53.

Friday 9/24. At Ashley's 1 day. Winans and Everett at Baldwin's.

Saturday 9/25. At Ashley's 1 day. Winans and Everett at Baldwin's. Paid Winans 10.00.

Sunday 9/26. At home. Rainy all day. No meeting.

Monday 9/27. At Ashley's 1 day. Everett and Jud at Baldwin's 1 day each. Cold, blustering day. Paid M. Kelso for tomatoes, 1.00.

Tuesday 9/28. At Ashley's 1 day on house. Winans and Everett at Baldwin's. Paid M. Griggs .54. J. Thompson had horse to Schenevus. Cotrell's family visited us.

Wednesday 9/29. At Ashley's 1 day. Everett and Winans at Baldwin's 1 day each. Paid S. E. Manzer .73. Heard Rev. Rowland preach in Baptist Church [in] evening.[83]

Thursday 9/30. At Ashley's 1 day. Winans and Everett at Baldwin's 1 day each.

OCTOBER

Friday 10/1. Took family to Schenevus Fair;[84] visited at W. E. Wright's. Paid J. K. Tyler, insurance,15.00;[85] Chase & Mills, 1.45; on the fair grounds, .60.

Saturday 10/2. At Ashley's 1 day. Everett helped Winans dig his potatoes 1 day. Received of W. O. Ashley, 50.00. Evening, Republican Caucus; delegates, J. K. Holmes, C. H. Bissell, D. Herdman.

Sunday 10/3. At home. Rainy all day; no meeting. Rainy all night.

Monday 10/4. Worked for M. J. Hubbard. Winans and Everett at Baldwin's 1 day; Everett, ¾ day. Rainy all day, very high water.

Tuesday 10/5. Worked for Ashley 1 day, laying floor. Everett and Winans at Baldwin's.

Wednesday 10/6. Worked for Ashley 8 hours; for Hubbard, 2 hours. Everett and Winans at Baldwin's 1 day each.

Thursday 10/7. At Ashley's 1 day. Winans and Everett at Baldwin's 1 day each. Wm. Pank's wife's funeral at Methodist Episcopal Church.[86]

Friday 10/8. At Ashley's 1 day. Everett and Jud at Baldwin's 1 day each. Arthur got in load corn; [we?] husked [it] in evening.

Saturday 10/9. Worked for Ashley 1 day [plus] 2 hours. Winans and Everett at Baldwin's 1 day each. Settled with Winans. Paid him cash, 13.00; all settled up to Baldwin's job.[87]

Sunday 10/10. Rainy day as usual. I attended Congregational Church; heard Rev. Moure.

Monday 10/11. At Ashley's 1 day. Winans and Everett at Baldwin's 1 day each. Baldwin worked for me on well 1 day.

Tuesday 10/12. I worked for W. T. Marks on water wheel 1 day.[88] Baldwin and Everett on well. Winans at Baldwin's. District School meeting; I am elected trustee.[89]

Wednesday 10/13. At Ashley's 1 day. Everett husked corn for me. Paid D. Salisbury school tax, 1.12.[90] Winans at Baldwin's.

Thursday 10/14. At Ashley's 1 day. Baldwin, Everett, and A. Cossart on well for me.

Friday 10/15. At Ashley's 1 day. Baldwin, Everett, and Cossart on well. Paid R. Hall, flour, 10.00; A. Cossart, work, 3.00.

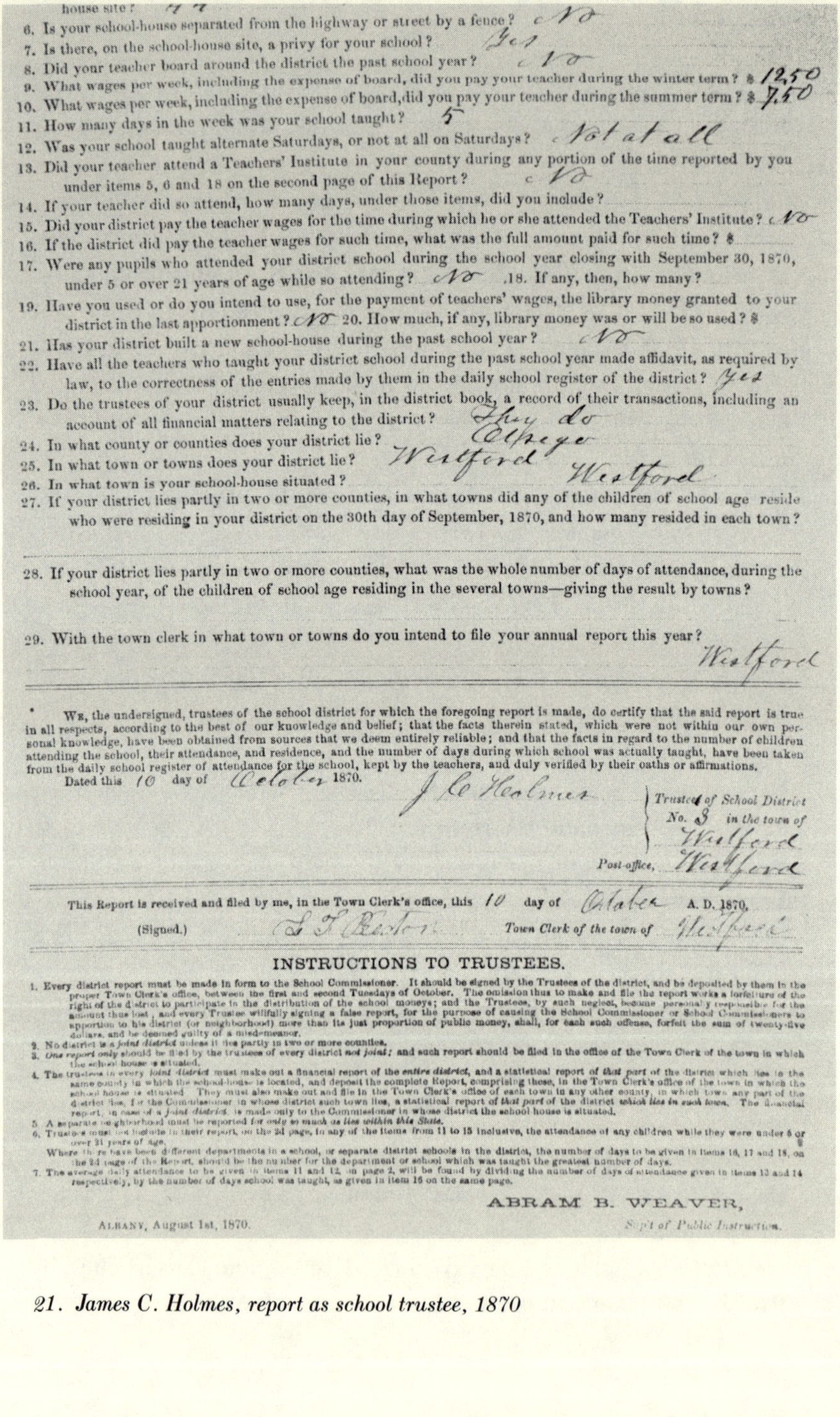

house site? ✓✓

6. Is your school-house separated from the highway or street by a fence? No
7. Is there, on the school-house site, a privy for your school? Yes
8. Did your teacher board around the district the past school year? No
9. What wages per week, including the expense of board, did you pay your teacher during the winter term? $ 12.50
10. What wages per week, including the expense of board, did you pay your teacher during the summer term? $ 7.50
11. How many days in the week was your school taught? 5
12. Was your school taught alternate Saturdays, or not at all on Saturdays? Not at all
13. Did your teacher attend a Teachers' Institute in your county during any portion of the time reported by you under items 5, 6 and 18 on the second page of this Report? No
14. If your teacher did so attend, how many days, under those items, did you include?
15. Did your district pay the teacher wages for the time during which he or she attended the Teachers' Institute? No
16. If the district did pay the teacher wages for such time, what was the full amount paid for such time? $
17. Were any pupils who attended your district school during the school year closing with September 30, 1870, under 5 or over 21 years of age while so attending? No 18. If any, then, how many?
19. Have you used or do you intend to use, for the payment of teachers' wages, the library money granted to your district in the last apportionment? No 20. How much, if any, library money was or will be so used? $
21. Has your district built a new school-house during the past school year? No
22. Have all the teachers who taught your district school during the past school year made affidavit, as required by law, to the correctness of the entries made by them in the daily school register of the district? Yes
23. Do the trustees of your district usually keep, in the district book, a record of their transactions, including an account of all financial matters relating to the district? They do
24. In what county or counties does your district lie? Otsego
25. In what town or towns does your district lie? Westford
26. In what town is your school-house situated? Westford
27. If your district lies partly in two or more counties, in what towns did any of the children of school age reside who were residing in your district on the 30th day of September, 1870, and how many resided in each town?
28. If your district lies partly in two or more counties, what was the whole number of days of attendance, during the school year, of the children of school age residing in the several towns—giving the result by towns?
29. With the town clerk in what town or towns do you intend to file your annual report this year? Westford

We, the undersigned, trustees of the school district for which the foregoing report is made, do certify that the said report is true in all respects, according to the best of our knowledge and belief; that the facts therein stated, which were not within our own personal knowledge, have been obtained from sources that we deem entirely reliable; and that the facts in regard to the number of children attending the school, their attendance, and residence, and the number of days during which school was actually taught, have been taken from the daily school register of attendance for the school, kept by the teachers, and duly verified by their oaths or affirmations.

Dated this 10 day of October 1870.

J C Holmes — *Trustee of School District No.* 3 *in the town of* Westford

Post-office, Westford

This Report is received and filed by me, in the Town Clerk's office, this 10 day of October A. D. 1870.

(Signed.) L. T. Keeton *Town Clerk of the town of* Westford

INSTRUCTIONS TO TRUSTEES.

1. Every district report must be made in form to the School Commissioner. It should be signed by the Trustees of the district, and be deposited by them in the proper Town Clerk's office, between the first and second Tuesdays of October. The omission thus to make and file the report works a forfeiture of the right of the district to participate in the distribution of the school moneys; and the Trustees, by such neglect, become personally responsible for the amount thus lost, and every Trustee willfully signing a false report, for the purpose of causing the School Commissioner or School Commissioners to apportion to his district (or neighborhood) more than its just proportion of public money, shall, for each such offense, forfeit the sum of twenty-five dollars, and be deemed guilty of a misdemeanor.
2. No district is a *joint district* unless it lies partly in two or more counties.
3. *One report only* should be filed by the trustees of every district *not joint;* and such report should be filed in the office of the Town Clerk of the town in which the school house is situated.
4. The trustees in every *joint district* must make out a financial report of the *entire district,* and a statistical report *of that part* of the district which lies in the same county in which the school-house is located, and deposit the complete Report, comprising these, in the Town Clerk's office of the town in which the school house is situated. They must also make out and file in the Town Clerk's office of each town in any other county, in which town any part of the district lies, for the Commissioner in whose district such town lies, a statistical report of *that part* of the district *which lies in such town.* The financial report, in case of a *joint district,* is made only to the Commissioner in whose district the school house is situated.
5. A separate neighborhood must be reported *for only so much as lies within this State.*
6. Trustees must not include in their report, on the 2d page, in any of the items from 11 to 15 inclusive, the attendance of any children while they were under 5 or over 21 years of age.

Where there have been different departments in a school, or separate district schools in the district, the number of days to be given in items 16, 17 and 18, on the 2d page of the Report, should be the number for the department or school which was taught the greatest number of days.

7. The average daily attendance to be given in items 11 and 12, on page 2, will be found by dividing the number of days of attendance given in items 13 and 14 respectively, by the number of days school was taught, as given in item 16 on the same page.

ABRAM B. WEAVER,

ALBANY, August 1st, 1870. *Sup't of Public Instruction.*

21. James C. Holmes, report as school trustee, 1870

22. Fanny Holmes Skinner and James B. Skinner, ca. 1885, and the Skinner house, ca. 1890, Westford

Saturday 10/16. Self and Everett at Ashley's 1 day each on house. Settled with Everett. Paid him 20.25; balance due him, 67.00. I lost [to him] 2½ hours.[91]

Sunday 10/17. At Methodist Episcopal Church; heard Fox talk.

Monday 10/18. At Ashley's 1 day. Winans at J. B. Skinner's.

Tuesday 10/19. At Ashley's ¾ day. Winans at J. B. Skinner's. Received of Mel Griggs, 20.00.[92] School meeting.

Wednesday 10/20. At Ashley's 1 day. Winans at J. B. Skinner's. Paid Everett Herdman 20.00 [and] sold him a cow [for] 40.00.[93]

Thursday 10/21. At Ashley's 1 day. Winans at Skinner's 1 day.

Friday 10/22. At Ashley's 1 day. Winans at Skinner's 1 day.

Saturday 10/23. At Ashley's 1 day. Rainy and snowy all day.

Sunday 10/24. At home. Called at J. K. Holmes's and W. T. Marks's.

Monday 10/25. Worked on cow stable for self; Winans helped. Received of Mrs. Alen, 10.00 for hay.

Tuesday 10/26. At Ashley's 1 day on house. Winans and Arthur went to Schenevus. Received of Waterman, by Winans, 5.00; Winans used 2.25 [of it].

Wednesday 10/27. At Ashley's ½ day. P.M., Winans and I went to Schenevus to look at apple trees.

Thursday 10/28. At home building cow stable. Bought 250 feet boards of Ashley [@] 1.20 per c [i.e., per hundred feet]. Winans worked 1 day. Cold, snowy day.

Friday 10/29. At home on cow stable ½ day; Winans helped. P.M., we went to Schenevus after trees. Paid E. C. Frost 15.00;[94] S. Gurney, 2.45. Received of S. E. Manzer, 15.00. Paid Winans 14.85.

Saturday 10/30. At home on cow stable. Winans had horse ½ day drawing wood from mill. Cold, windy day.

Sunday 10/31. Attended Methodist Episcopal Church; heard Rev Fox.

NOVEMBER

Monday 11/1. Worked for Ashley 1 day on pantry shelves and cupboard. Revilo Holmes commenced our school for 2.50 per day and board [to be provided by] self.

Tuesday 11/2. Worked for Ashley 1 day on pantry and base. Received of J. B. Skinner, 10.00. Paid F. P. Draper 5.00; paid W. H. Tyler 5.00. Received bill from Barnes & Fox, 158.20.

Wednesday 11/3. Worked for Ashley 1 day; put down all the base in chamber. Fine day.

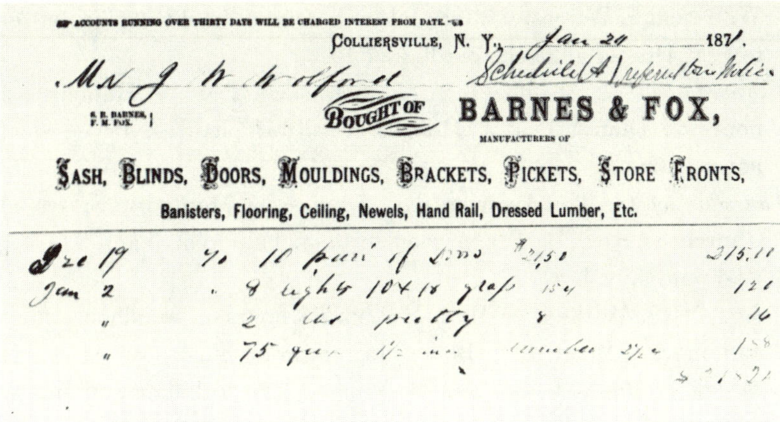

ACCOUNTS RUNNING OVER THIRTY DAYS WILL BE CHARGED INTEREST FROM DATE.

COLLIERSVILLE, N. Y., Jan 20 1871.

Mr J W Wolford

S. R. BARNES, F. M. FOX.

BOUGHT OF BARNES & FOX,

MANUFACTURERS OF

SASH, BLINDS, DOORS, MOULDINGS, BRACKETS, PICKETS, STORE FRONTS,

Banisters, Flooring, Ceiling, Newels, Hand Rail, Dressed Lumber, Etc.

23. *Barnes & Fox billhead, 1871*

Thursday 11/4. At Ashley's 9 hours. Town board met and did bounty claims; 312.00 [due] to me.[95]

Friday 11/5. At Ashley's 9½ hours.

Saturday 11/6. At Ashley's 1 day; finished.

Sunday 11/7. At home.

Monday 11/8. At M. J. Hubbard's hanging doors 1 day. Ashley plastering his house.

Tuesday 11/9. At Hubbard's ½ day. P.M., at home drawing wood. Called at Horace's in evening; bought 13 lbs. mutton of him, 1.30.

Wednesday 11/10. At Hubbard's 1 day on closet shelves and doors.[96] Bought 100 lbs. beef of Dan Brownell.

Thursday 11/11. At Hubbard's 1 day; put in window springs, hung and trimmed front door. Received of A. Snyder for Mrs. Alen, 14.75 to pay for blinds and doors.[97]

Friday 11/12. At home on cow stable and barn; built stall for horse, battened cow stable. Paid M. Griggs 20.00 borrow[ed] money.[98]

Saturday 11/13. At home drawing wood; Arthur helped. Got 8 loads with one horse. Bought 18½ lbs. butter of W. H. Tyler to fill firking [*sic*].

Sunday 11/14. At Methodist Episcopal Church; heard Elder [Fox].

Monday 11/15. At home. Took corn to mill and shelled it.

Tuesday 11/16. At home in A.M. At Hubbard's ½ day. Filed 2 saws.

Wife went to Worcester after Mother Shelland.[99] Sold 106 lbs. butter to F. P. Draper at 35 cts. per lb., 37.10.

Wednesday 11/17. At home in A.M.; filed saws. P.M., at Hubbard's 3 hours on pantry. Paid J. Holmes 5.00 [for] pictures. Pryne shod horse, 1.00.

Thursday 11/18. Thanksgiving day. Services at Methodist Episcopal Church by Rev. H. Fox; a good sermon. Paid Wm. Pank 2.00 for load pumpking [*sic*].

Friday 11/19. Worked for M. J. Hubbard 8 hours on woodhouse floor and bench. Attended Dr. Bolles's lecture in evening.

Saturday 11/20. At home. Paid J. Winans 5.70 to balance accounts. Attended Dr. Bolles's lecture on digestion; bought 2 course-tickets, 1.50. [Paid] M. Griggs .25.

Sunday 11/21. A.M., Baptist Church; heard Elder Rowland. Evening, heard Dr. Bolles lecture on temperance; the best I ever heard.

Monday 11/22. At home drawing wood. Fine day. Evening, heard Dr. Bolles lecture on the circulation of the blood and tobacco.

Tuesday 11/23. Worked for Hubbard 8 hours in house. Snow 3 inches deep. Evening, attended Dr. Bolles's lecture on the lungs.

Wednesday 11/24. At M. J. Hubbard's 8 hours. Evening, attended Dr. Bolles's lecture on the nerves.

Thursday 11/25. At M. J. Hubbard's 6 hours. Fine day. Evening, at home; Mary and Nettie at lecture.

Friday 11/26. At Hubbard's 8 hours. Attended lecture in evening.

Saturday 11/27. Worked for M. J. Hubbard 8 hours on cupboards. Evening, at home.

Sunday 11/28. At Methodist Episcopal Church. Heard Dr. Bolles in evening on "Trifles"; a good lecture.

Monday 11/29. At Hubbard's 8 hours.

Tuesday 11/30. At Hubbard's 8 hours. Rainy and thawy.

DECEMBER

Wednesday 12/1. At Hubbard's 6 hours. Filed saws for Pank, Z. Skinner, and J. M. Holmes. Received of Mrs. Alen, 15.00; W. Pank's filing, .30; F. P. Draper, 35.40.[100]

24. District School No. 3, ca. 1900, Westford

Thursday 12/2. At home sawing wood in woods. D. Cummings helped me 5 hours.[101] Filed saws for Tyler and Winner. Settled and paid E. R. Treet for beef, 1.14 + .75 + .75 [=] 2.64.

Friday 12/3. At home sawing wood in P.M. D. Cummings helped 4 hours. Filed saw for R. Pank and self. Sent check, 50.00, to Barnes & Fox, Colliersville.

Saturday 12/4. At home butchering hogs; Tucker helped. Lent W. T. Marks 8 lbs. spare ribs. W. E. Wright with us to dinner.

Sunday 12/5. At home. Cold and stormy.[102]

Monday 12/6. At home. Filed 3 saws for W. T. Marks. P.M., attended old Mr. Rose's funeral;[103] heard Elder Powers of Worcester. Snowed all day.

Tuesday 12/7. At home drawing wood from woods. Got 6 loads.

Wednesday 12/8. Drawing wood at home. Fine, cold day. Sleighing. Took Mary to Horace Holmes's.

Thursday 12/9. At home drawing wood for self; 8 loads. Fine day.

Friday 12/10. At home drawing wood. Sister Sarah [Wright] and child with us over night.[104] Bought of S. E. Manzer, 6 lbs. sugar [@] .16½; 5 lbs. salt of M. Griggs, 2½ yds. cotton flannel, 2/–.[105]

Saturday 12/11. At home drawing wood; Arthur helped saw. Mary up to Markses'. A. B. Chapman of Illinois in town.

Sunday 12/12. Attended Methodist Episcopal Church with A. B. Chapman and W. E. Wright; heard Rev. H. Fox. Called at J. M. Holmes's. Chapman and wife with us.[106]

Monday 12/13. At home visiting with Chapman and wife; [they are] stopping with us. Called at Horace's in P.M.

Tuesday 12/14. At home. Chapman helped me saw birch tree in back woods in P.M. Bought of H. W. Holmes, 150 lbs. beef at .08.

Wednesday 12/15. Went to Colliersville via of [*sic*] Schenevus; Mary with me to Chaseville. Settled and paid Barnes & Fox in full, 43.28. Hotel, .75; Schenevus, Bennet's shop, 6.82.

Thursday 12/16. At home. Settled with R. S. Hall at mill; paid him 4.19. Bought clothes and trimmings of W. H. Tyler, about 30.00 worth. Visited District School No. 3.[107]

Friday 12/17. At home splitting wood. Paid Erasmas Snyder for 5 barrels plaster and lime, 8.00; M. Griggs, goods, 1.00.

Saturday 12/18. At home in A.M. sawing wood. P.M., filed 2 saws. Received of W. H. Platner, Supervisor, 100.00.[108] Paid my tax for this year, 55.55; S. E. Manzer, goods, .81.

Sunday 12/19. At home all day. Blustering snow storm.

Monday 12/20. At home. Pack[ed] away wagon; got out cutter. Settled with M. Griggs, paid 14.75; paid F. B. Wright for 2 lbs. wool, 1.10.

Tuesday 12/21. At home. Made door for Ashley. Fitted and hung the same. Snow enough for good sleighing.

Wednesday 12/22. At home. Rainy all day; snow left. Settled with M. J. Hubbard; balance due me, 84.55, to be paid this month.

Thursday 12/23. At home. Put up stove in shop. Filed 2 saws for H. W. Holmes. Settled with R. S. Hall for 65 bunches lath and 1,065 feet lumber, about 40.00.

Friday 12/24. Worked in shop 3 hours for M. Griggs on flour box. Paid R. S. Hall, on set[tlement], 12.88; paid W. H. Tyler [on settlement] 42.00. Sent to D. Chester, 50.00.[109] Filed 2 saws; received 1.10.

Saturday 12/25. Attended Christmas at J. B. Skinner's; all the Holmeses [were] present except Clara and Emily. I gave Aunt Betsey 5.00.

Sunday 12/26. At home. Rainy all day.

Monday 12/27. At home. In shop ½ day. Arthur and I sawed wood in P.M. Received D. Chester's note [for] 50.00.

Tuesday 12/28. Worked in shop for M. Griggs 6 hours. Rainy all day. Received of M. Waterman ½ gal. oil, .60, [and] cash, 5.40.[110]

Wednesday 12/29. Attended H. Tipple's funeral in P.M.[111] Cut wood for self in A.M. Evening, meeting in Methodist Episcopal Church.

Thursday 12/30. Worked in shop; filed saws for Z. Skinner and F. B. Wright. Received of Z. S[kinner], .50.

Friday 12/31. Arthur and I sawed wood at door and in woods. I filed saw in morning for J. B. Skinner; made board for saw. Attended meeting at Methodist Episcopal Church; heard Fox.

3 The Major Jobs of 1869

During 1869, James Holmes carried on, in person or through his assistants, a total of six major jobs, besides building a new shop, a cow stable, and an outhouse for himself. The major jobs were roughly sequential, although they partly overlapped and a given job was interrupted on occasion and then returned to later. First came the house of Judson Winans, begun sometime in late 1868 and finished—as far as Holmes's contribution was concerned—early in the new year. Next, Holmes turned to his own new house, on which he and a number of other workers labored until June and, in the case of a couple of the men, July and even August. After leaving his own job, Holmes spent all of July and the bulk of August at work on the Hubbard house; then, to the beginning of November, he turned to the Ashley job on his own while Winans pursued other work, first with Herdman at German Baldwin's farm and then alone at James and Fanny Skinner's. At year's end, Holmes did finish work at the Hubbard and Ashley sites.

A detailed analysis of the first four jobs follows, with a final section on the others. Each analysis is preceded by an outline of the labor performed on a given job and a list of the workers known to have been engaged on it.

THE JUDSON WINANS HOUSE

Nov.–Dec.	1868	Frame erected and sided
Jan. 6–9	1869	Roofing laid

Jan. 14	Piazza built
Jan. 18–19	Stairs erected
Jan. 20–Feb. 13	Windows and doors cased
Feb. 16–24	Finish work

Workers: James C. Holmes, Judson Winans, William E. Wright

Holmes started the new year finishing up on Winans's house. It probably was located north of the village on a lot along Elk Creek road; when I checked in 1983, the lot was vacant. Holmes must have prepared and raised the frame for this house before going to Albany late in December of 1868. It is possible to estimate how long he spent on these aspects of the job. Holmes was still in Westford on December 5, when he attended the monthly church meeting. When he later totaled his 1869 hours on the job (memorandum 4), he added to that verifiable figure another, unexplained, eighty-three hours. If, as seems likely, this figure represented his 1868 hours for Winans, Holmes probably had remained in the village for at least two weeks after December 5. It took him almost three weeks in January, the diary shows, to work an equivalent number of hours for Winans.

Shortly after he returned to Westford on January 2, Holmes was boarding the roof—typically the roof boards went on after the walls were sided—and then shingling it. Holmes next put up the stove and hung a door (probably the front one) before erecting the piazza. When that was finished, he moved inside to build the stairs, case the doors and windows and put on their stops, lay the sills, install shelves in the pantry and the clothes press, install the columns (these may well have been outside), build a set of cellar stairs, and finally install the banisters and handrail on the main set. It would appear that Winans, perhaps with the aid of Holmes's brother-in-law William E. Wright (Milton Wright's son, noted as aiding Winans on January 6), had done a good deal: siding the frame after it was erected, probably installing the windows and most of the doors (Holmes mentions casing them but not installing them, although this may be shorthand for both operations), laying the floor while Holmes was working on the roof, and either plastering the house or having someone else do it after Holmes had left the job on February 24, probably once the weather warmed up. After the plastering was finished, the base moldings (which Holmes does not mention) would have to have been laid. Apparently the stove put up in January was vented by a simple pipe rather than a chimney (Holmes's house, it's clear, was to have a proper chimney)—or if Winans was to have a chimney it would be erected in warmer weather.

THE JAMES C. HOLMES HOUSE

Jan. 6–Feb. 20	1869	Lumber and shingles readied; millwork ordered
Feb. 10–Mar. 2		Stone quarried and hauled
Mar. 4–Apr. 10		Windows prepared
Apr. 14–24		Frame hewn
Apr. 30–May 7		Cellar excavated
May 10–27		Foundation laid up
May 28–29		Frame raised
June 2–7		Siding put up
June 8–11		Roofing laid
June 12–16		Flooring laid; exterior painted
June 17		Bench moved inside; stairs built
June 18–21		Cellar plastered
June 19–21		Partitions built
June 22–28		Base and casings installed
June 26–July 9		Lathing nailed up
July 26–28		Chimney built
Aug. 2–7		House plastered

Workers: James C. Holmes, Horace Holmes, Arthur Holmes, Mary Holmes, Judson Winans, Everett Herdman, German Baldwin, Augustus Baldwin, Charles Tyler, John Goodenough, Denton Salisbury, William O. Ashley, Carlos Ashley, Myron Hubbard

When Holmes left the Winans job, he did not immediately begin work at another site. Instead he began to prepare the shopwork items for his own new house. Already in January he had worked on the doors, and in early February he had begun sawing out the casing and base moldings. In March he began putting in full days, first on the frames, then on the casings and base for the sash he had ordered on February 1 and February 5 from the millwork firm of Barnes & Fox in Colliersville, a village on the new rail line about fifteen miles from Westford. Holmes was doing this work in his shop, in a barn he was renting from the widow of a cabinetmaker in the village, and perhaps in part of his present house. (His note on March 10, "moved into the north part of the house," cannot refer to the new house—no part of which was standing at this time—and may refer to Holmes's use of the north end of his old dwelling as a work area.) For the most part, Holmes directed his attention now to the house frame. He was building hewn timber houses, not modern balloon frame ones, and all the members needed for the raising, which was to take place on May 29, had to be carefully prepared in advance, either by him or (as may have been the case in the other jobs) by someone else. But Holmes also was doing other things at the same time. Apparently his present shop was not adequate to his needs. It was at the least remote from the site of his new house, and he would need a new one there once he moved anyway. Late in April he thus began building the new shop (fig. 25). He hewed its frame from April 26 through April 29 and must have erected and finished it off from May 10 through May 18, since it was "ready to move into" on the latter date; on the 19th Holmes in fact did move in. Here, at the edge of his farmland across the Worcester road from his house site and thus about five hundred feet southeast of the village, he was to have his base of operations over the remaining months of the year.

Holmes had assembled his new farm from three separate parcels purchased in 1868. There was a 1½-acre piece which he bought from his sister Sarah and her husband, William E. Wright, for $125 in April. This was a triangular parcel lying on the south side of the road.[1] A steep drop-off at the rear overlooked Elk Creek as it switched to the east after passing under the road closer to the village. Straddling the stream below the parcel was Elk Creek Mills, being run in 1869 by the Scotsman Robert S. Hall but previously owned by Myron J. Hubbard. Holmes's two other parcels lay side-by-side across the road. The first, containing 5½ acres, he had bought later in 1868 from Phoebe D. (Dolly) Bentley, the wife of Baptist deacon Russell Bentley and thus someone long known to Holmes. He agreed to pay Bentley $400 for this

25. James C. Holmes, new carpentry shop, 1869, Westford

parcel. His payment of $28 in interest to her on April 1, 1869, as recorded in the diary, evidently means—it is a round 7 percent of the principal—that he had yet to discharge any of the debt. No mortgage was ever filed in this case, but Bentley in effect was holding one for Holmes. This land lay like a wedge between the road and the base of the big hill that shouldered the village on its east flank.

The third parcel, lying between this wedge and the village proper, contained 45 acres. Holmes purchased it in September 1868 from Sarah and Will Wright for $2,000, with the stipulation that they were to pay the interest due on two old mortgages through April 1 of 1869 and that Holmes was then to assume those mortgages. This was the biggest of the parcels and the one most suited to agricultural purposes. A narrow tract, it was bordered on the west by Elk Creek, and here lay the fringe of meadowland noted in the agricultural census. On the east it lay beneath the hill running north along the valley, a hill that passed behind Calvin and Horace Holmes's farms, above the village, before passing out of sight altogether. But much of James's land between the creek and the hill in fact was relatively flat (fig. 26). When Holmes bought it,

26. The Holmes farm, 1869–1895, Westford

the parcel probably was one of the best pieces of farmland close to the village.

His farmstead was divided by the Worcester road, which plunged down quickly from Westford's main street toward the Elk Creek crossing. The road ran a bit north of the house Holmes was to build for William O. Ashley in 1869, then passed miller Robert S. Hall's (on the right before the creek), crossed the stream, and climbed past Denton Salisbury's. As it rose farther it curved to the south and then straightened and leveled out briefly (fig. 27) before swinging back to the left and rising steeply again. It was in this flat interval that Holmes located his farm. He built his house on the smallest of the parcels, that south of the road. His shop he erected slightly west of the house and across the road, at the near edge of the largest parcel. His barn, either the one he built in late 1869 or a later structure, was to stand east of the shop, just under the hill on the 5½-acre parcel.

The shop Holmes built was a simple 1½-story gabled structure with a 1-story shed addition on its right side, the whole about twenty by twenty-five feet. To its west the barn, since torn down, contained (so

27. Worcester road near the Holmes farm, Westford

recalled the grandson of a former owner of the farm)[2] about ten stalls. It was oblong, with its short end to the road. Between it and the shop evidently was Holmes's farmyard. Here a later owner once erected a small lumber mill and a hoghouse, neither standing when I first visited the site in 1983. On the far side of the shop, toward the village, Holmes apparently planted the apple trees he purchased in Schenevus late in 1869. Remnants of the orchard today line the edges of the small field there. Farther back on the land, about on line with the shop, is a second smaller outbuilding (fig. 28), seemingly of the same period as the shop.

The house itself as Holmes erected it was a 1½ story vernacular Greek Revival structure, later extended by a 1-story shed addition on the rear which recently has been extended again (figs. 29, 30, and 31). The front door was at the extreme left corner. This gave entrance to a small hall from which a narrow staircase led directly upstairs. Through a doorway to the right, the hall also gave entrance to the front room of the main floor, which occupied the whole front. Its outer walls, pierced by two windows on the front and one on the far side, were plastered; the partition at the rear that divided it from the back room was made of vertical boards. The back room later had its far right corner enclosed so as to provide a bathroom. On the opposite (east) outer wall of this back room there was from the beginning a doorway, framed by a pair of windows, that led out to a small porch covered by a single-pitched roof. Evidently this back room was the original kitchen, but at present the rear shed addition serves this purpose. It rests on its own separate foundation, as does the recent second extension at its rear. The basement runs only under the house proper, but the entrance to it—originally a bulkhead against the rear wall of the house—now heads down from the center of the shed, with a small pantry lodged above it. One faces toward the front of the house in descending the stairs. At the far right corner of the shed additions, with access from the back room of the house proper, is a small downstairs bedroom. I did not inspect the upstairs of the house, but interviewees described its layout sufficiently well for me to indicate it here. The largest room runs across the front, directly over the front room downstairs. Behind it is a smaller second bedroom. At the head of the stairs is a third, even smaller room to which access is had from the rear bedroom.

28. James C. Holmes, rear outbuilding, ca. 1869, Westford

29. James C. Holmes, Holmes house, 1869, northeast corner, Westford

30. Holmes house, east side

31. Holmes house, west side

The trim on the Holmes house is extremely simple (figs. 32 and 33). Flat corner boards, water tables, and cornices frame the 5½-inch clapboard siding. Similar boards case the windows, all of which probably were 6-over-6 originally (except for the 6-light fixed sash on the sidewalls of the second floor), although now a number have been replaced with single-pane sash. The village provides finer examples of the Greek Revival style Holmes chose for his house. Especially notable is the Mel Griggs store (fig. 16), with its four massive Ionic columns, the pediment that rests on them, and a full cornice that runs back along the sides. Holmes, however, was building within the vernacular rather than the polite range, and what he produced here was a serviceable but modest house that participated in current fashion only minimally. The debt of his dwelling to the Classic Revival movement of the earlier decades of the century is shown by its orientation and its plan rather than by its detailing.

Holmes's diary records with great specificity the steps leading to the completion of the house. As noted above, in the case of other houses, Holmes served as builder rather than contractor and thus was responsible only for his own labor and that of his helpers. In his own house, however, he was the contractor as well and thus was responsible for all the steps. The diary supplies much detail, for instance, regarding the preparation of the site and the building of the foundation. We know that Holmes and Everett Herdman began digging the cellar hole on April 30, German Baldwin having brought a team of horses and (apparently) a scraper to help. Excavation continued through May 7 or May 8, and on May 10 William O. Ashley and his helper or helpers began laying up the foundation. Long before this, on February 10, German Baldwin and his son Augustus had started work in the local quarry,[3] and Holmes visited them there on the next day to check on their work. He also joined his brother Horace, his son Arthur, and Charles Tyler (who may have owned the quarry) in picking—that is, prying loose—and hauling the many loads of stone needed for the house foundation and perhaps the well. This heavy task occupied a number of days from February 17 until March 2. Then again from May 10 to May 19, Holmes and his son Arthur hauled more stone as the masons began their work. Holmes counted thirty wagon loads in his

32. Holmes house, front trim

33. Holmes house, corner trim

"House for Self" (memorandum 6) summary of costs, although the daily record through March 2 yields a total of at least thirty-eight, and that leaves five other days out of consideration.

And stone was only the first commodity to be hauled. Holmes and others also had to haul large amounts of sand and lime and plaster, as well as such bulky supplies as shingles, brick for the chimney, lath for the plasterers (though this came from his next-door neighbor Robert S. Hall, the miller)—not to mention the house frame itself, the siding, the roof boards, and the windows (some of which Holmes picked up in Schenevus and brought up the valley on April 22). Conditions were not always ideal for this labor: for instance, Holmes brought one load of lime home from Schenevus "in the rain" on May 15. And whatever the weather on a given day, all this labor was grinding and intensive: one notes that between February 1 and June 11, for example, the diary contains more than a dozen entries relative to procuring, hauling, stacking, applying, and settling for so lowly a building item as shingles.

The foundation was finished on May 27. The next day, as Denton Salisbury graded the soil around the site, Holmes, Herdman, and probably Winans put the sill on top of the walls. Then, on May 29, the three men—with Salisbury's help—raised and secured the frame. When they returned to the job on Monday, May 31, they probably continued to work on the main framing members—perhaps on the roof timbers—and on that day and the next they also would have been nailing up the studs and laying the joists. On June 2 they began siding the frame, a task finished on June 7. On the next day, they were up on the roof frame, boarding it over and then laying the wood shingles; this took them until June 11. The next step was to lay the flooring, a job that Holmes and Herdman began on June 12 and apparently finished on June 16. They had only a small amount of help from Winans, who spent part of June 12 and all of June 15 and June 16 painting the exterior of the house.

At this time, the house was a completed shell. When Holmes moved his tools inside on June 17, a step that marked just this point in most jobs, his immediate concern was to build the main stairs and erect the interior partitions. The stairs, as noted earlier, were a simple set leading directly up from the small foyer. The partitions were built of boards and were not to be plastered. The (uninspected) upstairs rooms apparently had no plaster on any walls: how else could Mary Holmes have "painted [the] chamber" on June 24, before any plastering was reported in the diary and long before fresh plaster could have cured anyway?

Holmes also began to install the doors and the windows at this time, and then to trim those millwork items. These tasks, along with some lathing done by Winans on June 30, occupied the three men through the rest of June. But even before they left the job on the last day of the month, part of Ashley's crew had returned to the house to begin their next step in the process. Had they in the meantime been excavating for the Hubbard job and laying its foundation? Holmes made cellar window frames for Hubbard's house on June 10; these would be needed as the foundation neared its top. And since Holmes was to erect the Hubbard frame from July 8 to July 10, the foundation clearly had to have been finished by then at the latest.

However that may have been, on June 18 Holmes had taken time off from his carpentry tasks in order to mix mortar, with his son Arthur, for his own cellar wall. Either Carlos Ashley already was back at the site then and they were mixing for him, or they were premixing the material to ready it for his recorded return on June 21, when he "plastered [the] cellar." On June 26, young Ashley and his father's helper John began lathing the first-floor frame, a task that occupied them full-time until June 29 and then again on July 8 and July 9. From the latter date until the end of July, Carlos and John apparently were not on the Holmes job. One may speculate here, rather as above, that they were at work on the foundation of the Ashley house, which had to be ready for the erection of the frame by Holmes on August 27. If that was so, the Ashley job must have been in readiness by July 26, when William O. Ashley himself, aided by Myron J. Hubbard and young Carlos, first erected a chimney in Holmes's house (July 26 to July 28) and then plastered the first-floor walls lathed earlier by Carlos and John (August 2 through August 7). Some support is lent to this hypothesis by the fact that on June 29 Holmes made cellar window frames for Ashley's house.

It is possible that the Ashley foundation was already underway then. This was precisely the date, by the way, when Carlos Ashley and John left the Holmes job (not to return until July 8), so they may have begun the Ashley foundation even this early.

By August 7, the Holmes house was virtually finished. Holmes had built a backhouse (or privy) for it on May 22; the new well would be dug, and probably cased with stone, by German Baldwin in mid-October. Little else was needed for the house proper to be habitable, although nowhere in the diary does Holmes indicate that his family moved into it. If the three pictures he bought on August 14 were decorative items intended for the sitting room of the new dwelling, then perhaps by then the family already had moved in. That they had done so by June of the next year at the very latest is proven by the census taken then, for their position in the listing for Westford village clearly places them precisely there.

THE COL. MYRON J. HUBBARD HOUSE

June 10	1869	JCH makes cellar window frames (foundation in progress?)
July 1–7		Frame hewn
July 8–10		Frame raised
July 12–15		Siding put up
July 16		Wing cornice put up
July 16–29		Roofing laid; bench moved inside
July 30–Aug. 21		Flooring laid; base and casing installed; partitions built

Workers: James C. Holmes, Judson Winans, Everett Herdman, Myron J. Hubbard

Once Holmes himself stopped work on his new house on June 30 of 1869, he hardly had entered a more peaceful period. The very next day, he and his two regular helpers began to prepare Hubbard's frame. This task occupied them July 2 and July 3 as well. The Fourth of July fell on Sunday, so the three men took off Monday, July 5, as a holiday. (Holmes and Winans went to Schenevus together, but not to watch any celebration—the newspaper indicates there was none in that village in 1869.) They returned to the task on Tuesday, July 6. Winans was off again on July 7, so he could attend a meeting of the Worcester Baptist Association in nearby Clarksville, and then on July 8 the three were back together, raising the frame in the rain that afternoon. The house, built on a location that cannot be identified with certainty, was evidently (as the Ashley house also was to be) an upright-with-wing design; but Holmes's reference to the Hubbard wing in July does not specify where that structure stood in relation to the house proper, so it might have been located at the structure's rear. In any case, the whole Hubbard frame was up by Saturday, July 10.

During the next week, the three men each worked six full days (except that Everett was off Wednesday afternoon), evidently putting up the studs, laying the joists, and nailing the rafters in place. By July 16, the walls must have been enclosed and the roof boards put on as well, since on that day they were installing the wing's cornice and had begun to shingle its roof. Presumably, the shingling of that roof was finished sometime before Holmes noted on July 29 that the men had "finished shingling" altogether; that note in all likelihood refers to the roof of the main house. On the same day, however, Holmes recorded moving his bench and tools inside, which he could have done only if part of the flooring also had been laid by then, so the dates here are not entirely clear; it is conceivable that the shingling was done not all at once, or even in two spurts of activity, but rather intermittently, as the flooring also may have been done. After this point, the crew must have turned in earnest to the windows. Holmes noted on August 6 that "all hands [were] casing up." Over the next two weeks, until Holmes wrote on August 21 that the job was "finished for the present," they worked on the sash and the doors, as well as on the partitions. They also may have lathed the walls that were to be plastered, in their absence, over the next several weeks. Holmes was to return to the Hubbard house in October and November for his finish work, but it seems best to defer consideration of that period of labor until later.

THE WILLIAM O. ASHLEY HOUSE

June 29	1869	JCH makes cellar window frames (foundation in progress?)
Aug. 23–26		Frame hewn; sills put on
Aug. 27–28		Frame raised
Aug. 30–Oct. 2		Water table put on; siding put up; roofing laid
Oct. 5–27		Flooring laid; stairs built; base and casing installed; partitions built
Nov. 1–6		Finish work
Dec. 21		Door made and hung

Workers: James C. Holmes, Judson Winans, Everett Herdman, William O. Ashley

Once again Holmes was busily moving from one job to the next. August 21 was a Saturday. On the following Monday, he and his two helpers began to prepare the frame for William O. Ashley's house, an upright-with-wing structure to be erected on Ashley's vacant lot just south of the Episcopal church in Westford village (fig. 34). They spent three full days on it, plus much of a fourth, before they could put the sills on the top of the foundation walls on August 26. The next day they erected the frame, and on Saturday, August 28, they must have been studding the walls, laying joists, and putting up the rafters. On Monday, Winans put on the water table, after which the siding would be nailed up. Once that step was done, and perhaps the roof had been finished as well, Winans and Herdman left the Ashley site to begin work on another job. Left alone here, Holmes henceforth stopped noting in his diary precisely what he was accomplishing each day. As a result, his record becomes rather sparse. But a note on October 5, which indicates that he was then laying the floor, makes possible a comparison of the time spent up to this point with the more detailed records for the Hubbard job. This comparison in turn helps to suggest what had been accomplished in the interim. On the Hubbard house, the men had spent a total of approximately 650 man-hours from the preparation of the frame until they moved the bench inside. On the Ashley house, it took about 535 man-hours for Holmes to begin flooring over the joists.

34. James C. Holmes, William O. Ashley house, 1869, Westford

Hence it seems proper to infer that the procedures he followed while alone on the Ashley job were roughly the same as those followed on the preceding one by Holmes and his helpers.

Later in the Ashley job, however, Holmes seems to have put in much less time. From laying the floor on October 5 until he noted on November 1 that he was installing the pantry shelves and the cupboard, he had devoted only another 80 or so man-hours. And by the time he had finished on November 6, the total time he had put into the job since the flooring was laid was a mere 120 man-hours. This was only about 15 percent of the time put in on the Hubbard job from a similar point to its end. How to explain this wide discrepancy is not clear. Perhaps it is best to regard Holmes's "finished" on November 6 as implying "for the present." In fact Holmes did return in December to install a door he had made for the house. If more such work remained to be done in 1870, then the house simply had not been completed by year's end. There is evidence pointing either way on this issue. On the one hand, the fact that Ashley started plastering the house on November 8 suggests that all but some of the trim work had been finished by then. On the other hand, the fact that Holmes settled with Hubbard for that man's new house in 1869 but nowhere recorded settling with Ashley suggests that the latter's dwelling was not yet done. When jobs were finished, there clearly was a strong impetus for local residents to settle up as the year closed. Perhaps Holmes had managed to build the stairs and to install some or all of the windows but not to case them up or put down more base molding than that which he noted, on November 3, installing in the chamber. Or perhaps Ashley himself, and some of his workers, labored alongside Holmes on some of the interior tasks needed to bring the house near to completion as the year was ending. Whatever did happen, it seems virtually certain that Holmes on his own could not have finished the house within the time allotted.

THE HUBBARD HOUSE REVISITED

Oct. 4–6	1869	Finish work
Nov. 8–Dec. 1		Finish work

Dec. 15	Finish work
Dec. 22	Finish work

Worker: James C. Holmes

When he left Ashley's on November 6, Holmes returned to Hubbard's. He had spent some time even while on Ashley's job, on October 4 and October 6, off laboring at Hubbard's. But now he devoted sixteen days over a period of 3½ weeks to the finish work left undone before. He hung doors on November 8, worked on the closet shelves and doors on November 10, and installed window springs and hung the front door on November 11. On November 17 he worked on the pantry and on November 19 on the woodhouse (laying its floor and building a bench), only to go back into the house proper on November 23. Here he worked for another eight days, usually for six or eight hours per day, building cupboards (so he notes on November 27) and perhaps doing other similar tasks.

By December 22 the house must have been complete, at least so far as his labor was concerned, for on that date the two men settled their accounts. Hubbard had paid him $85.00 on December 15 and now agreed that he owed Holmes another $84.55. The latter sum was to be "due this month," but neither the diary nor the cash account shows that Hubbard discharged the debt then. Since Hubbard had already paid Holmes, in four earlier installments in August and September, a total of $100.00, the sum they agreed he owed Holmes for all of the latter's labor (and that of his helpers) on the house in 1869 was $269.55. That this was only for labor must be obvious merely from the relative size of the figure. But in fact Holmes kept an exact account of the days he and Winans and Herdman put in on the house, and on August 21 he even totaled those days. When these times are multiplied by daily rates applicable in each case (for Winans and Herdman, I have used, as discussed earlier, not $1.50 and $.75 but rather $2.00 and $1.00), the total comes to $264.75, less than $5.00 off the mark. If these calculations are accurate, then it seems inescapable that Hubbard supplied all the materials for the job, even the framing timbers, and that Holmes served only as the carpenter for it. A minor adjustment necessary to accommodate Hubbard's labor on the Holmes house (priced by Holmes at

$15.00) and the possibility (memorandum 7) that Holmes had ordered millwork items for Hubbard and charged him a 5 percent markup (or perhaps passed along his own 5 percent discount) might together alter the terms of this analysis, but they should not change its basic assumptions. Myron Hubbard was indeed his own contractor. So, too, we may assume, was William O. Ashley.

THE OTHER JOBS OF 1869

With the Winans house finished, his own house substantially completed, and the Hubbard house framed, covered, and roughed in inside—so things stood by late August—Holmes had taken his crew, as we have seen, to the Ashley job. But once the frame was raised there, Winans and Herdman left Holmes to begin another job for Holmes on their own. This was for the laborer German Baldwin. Already in 1868, according to the county atlas published then, Baldwin owned a house on what is now called Strong Hill Road north of the village, and he seems to have stayed in that location after 1869. It thus is not clear what kind of work Winans and Herdman did for him in 1869. In all likelihood, they did not build a new house, although this would have been possible on the site where his house was recorded as standing in 1868. Perhaps that one had been destroyed, or was to be replaced, or was to be joined nearby by another, a not-uncommon practice in the period. But Winans and Herdman alone could not have built Baldwin a new house, unless it were very small and very common. They put in, together, a total of about 580 man-hours, less than what was required to bring the Ashley house to the point where Holmes could start his interior work there. And even if Baldwin and his son Augustus were helping, four men of their collective talents probably could not have brought a house to its finished state. For that, someone like Holmes would have been required, someone able to build stairs (remember that Holmes built the stairs in the Winans house in January, and Winans was clearly the most talented of the four men on the Baldwin job), to do the casing and the base (although Winans probably could have done this, in fact Holmes did a good deal of it for Winans in February), to put in whatever cabinets and storage closets were needed (Holmes also did these things for Winans), and in general to execute the other finish work on the job.

Yet if it seems certain that Baldwin was not building another house and employing Winans (through Holmes) as his main carpenter, it also is certain that the job in hand was of major proportions. Holmes's note (memorandum 7), which indicates that he purchased "about 9 windows" for Baldwin in 1869, and his diary entry on September 7, which reports that Baldwin bought "two sticks of timber" from Holmes (these would have been framing members of some sort), together suggest that some substantial domestic work (rather than, say, a barn) was being undertaken. Perhaps one or both of the additions visible on Baldwin's house in 1983 were now being built. Whatever the precise nature of the job, that Winans was put in charge of it (as noted earlier) marked Holmes's growing confidence in his work. Similarly, that Holmes could handle jobs through his employees rather than in person suggests that his standing as a craftsman in Westford at this time was relatively high. If the more constricted job Winans was assigned to do for Holmes's sister Fanny Skinner and her husband, James, after he finished at Baldwin's did not redound much to his credit or to that of Holmes on this issue, the Baldwin job clearly did.

Late in October, as Holmes was about to leave Ashley's job and Winans had just finished Baldwin's and the Skinners', the two men began to work together on Holmes's new cow stable. He owned in 1869, the 1870 census reported, a total of six milch cows and one other cow or steer. The 1869 diary suggests, as well, that he regularly bought and sold cattle, and so he probably had in his possession at any given time even more animals than the average figure given in the census. Since all of them needed shelter (and care), it seems reasonable to assume that the stable Holmes erected in 1869 was of a reasonably large capacity. But beyond that assumption it is difficult to be specific. Relatively little labor on the structure was recorded by Holmes in the diary. He and Winans worked on it all day on October 25 and October 28, plus a half day on October 29. Holmes alone worked all day on October 30 but then did not record more labor on the structure until November 12, when he noted that he was "at home on cow stable and barn" and that he "built stall for horse, [and] battened cow stable." Taken together, these last comments may indicate that the barn formerly located on the east edge of Holmes's farmstead dated at least in part from 1869. But the diary entries are too sketchy for one to go further. Suffice it to say that as Holmes's extraordinarily active year was drawing to a close he still had much that needed doing, so much so that he started work on the stable even before he was done at Ashley's and he continued to work on it as he returned to Hubbard's. And Holmes once more now, as he had early in the year, was beginning to file woodsaws for his neighbors. For the time of tree-felling and wood-cutting was again on the land, and soon the snow would follow.

At last, as winter began to come on, James Holmes could see to his own domestic needs—sawing wood himself, butchering hogs, taking his corn to mill—but he also began to relax the heady pace of the past eleven months. One feels in the final entries of his diary an easing up as he visits with his sister Julia and her husband, Alpheus, just arrived from Illinois, or buys "clothes and trimmings of W. H. Tyler, about $30.00 worth," or stores his wagon and gets out his cutter a day after a "blustering snow storm," finding out the next day (he must have tested it) that there is "snow enough for good sleighing." And one senses, too, amid the settlings up with clients, employees, and local businessmen, and the settling down for another long winter in the New York hill country, that Holmes could look back on the year past and regard it as an especially full vindication of his success. He had not kept his diary in order to pass on a record of that year or his labor. But that it survives, like a shaving on the shop floor, allows us to sense by implication a good deal of the process by which he had achieved his ends in 1869.

PART TWO CONTEXT

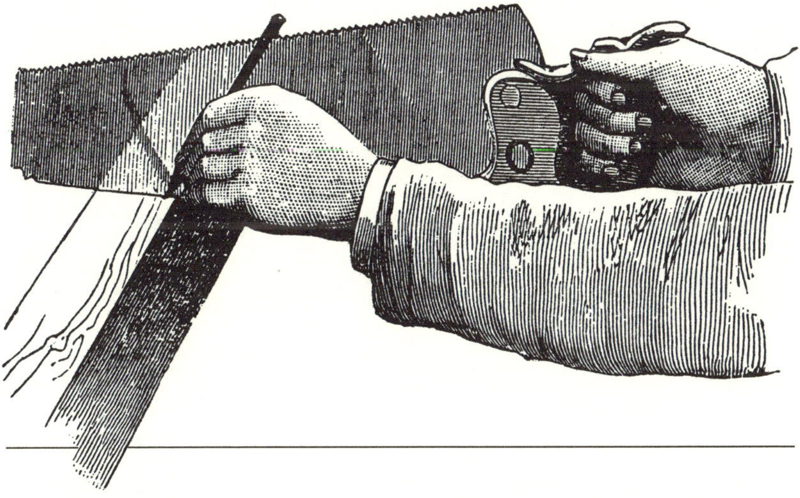

4 Carpentry in Otsego County in the Nineteenth Century

Although James C. Holmes well may have regarded his accomplishments in 1869 with a sense of pride, his success as a carpenter was both relative and temporary. The region in which he lived and worked in fact was economically hard-pressed, and for carpenters in particular it offered few permanent opportunities. Perhaps it was this constricting sense of possibility, rather than any longlasting commitment to agriculture, that led Holmes to redefine his occupation in the census schedules between 1855 and 1880.

The enduring problem here was demographic: a drain of the local populace, such as that seen in Holmes's own family, which left Otsego County as a whole with an almost flat population for the span of Holmes's life and beyond. The underlying causes of this drain were economic and topographic. Bypassed by the Erie Canal in the 1820s, and locked within the difficult embrace of its beautiful hills, most of Otsego County did not participate fully in the extraordinary growth that transformed nineteenth-century New York from a rather backward member of the new American union into the most populous and wealthiest state.[1] In common with the other counties south of the canal route, Otsego languished after reaching an early peak. In 1830, the year James Holmes was born, these counties had accounted for fully 29.1 percent of the state's population, the largest percentage of any of the state's regions; but by 1855, this region's share had fallen to 22 percent, and then to 18.4 percent in 1870. In 1892, three years before Holmes died, the share had fallen to 13.7 percent, and by 1925, the centennial of the opening of the canal, the share was a mere 9.4 percent. Not even the mountainous counties north of the canal route, even less promising for commerce and agriculture than the southern ones, experienced so large a drop in relative percentage. Every region outside of New York City and its suburbs, of course, showed percentage losses across this whole period. But the counties that touched Otsego just a few miles north of Westford, counties traversed by the canal and later by the state's major railroads, almost held their own, falling from 26.7 percent in 1830 to 20.1 percent in 1925. Given the extraordinary real growth in New York's population over the period, this relative fall actually constituted a sizable absolute increase.[2]

As close as it was to this more prosperous region, Otsego County, and especially its inland townships in the hilly eastern section where Westford lay, actually fell, in absolute terms, across this same period. The county's early high of 51,372 in 1830 was briefly surpassed in 1880, when 51,397 were counted, but the figure had fallen as low as 48,616 in the interim and by 1940 was at 46,082. Since the township of Oneonta in the southern part of the county was experiencing substantial growth in the decades following the opening of the Albany and Susquehanna Railroad in the late 1860s, pushing from a total of 1,902 in 1850 to totals of 8,018 in 1890 and 14,414 in 1940, the rest of the county obviously lost far more than its nominally flat or nearly flat totals across this period would imply. Otsego, excluding Oneonta township, that is, had a total population of 46,736 in 1850; by 1940, this figure had fallen to 31,668.

The population losses were heaviest in the area surrounding Westford. Decatur, lying to the east of Westford along the border with Schoharie County, fell from 927 in 1850, when it was the county's smallest township, to a mere 324 in 1940. In 1830, when Decatur had been the second smallest township, Westford was the fifth smallest. It then had an all-time high population of 1,645. By 1850, however, Westford had fallen to second smallest, with a population of 1,423, a drop of 13.5 percent. Its trend thereafter was generally downward, as table 1 shows.

That people were moving out of the county in large numbers is clear from these figures. Otsego County and neighboring Schoharie County in 1875 reported the state's highest percentages for inhabitants born

Table 1. Total Population, Westford Township, 1855–1940

Year	Population	Year	Population	Year	Population
1855	1,371	1890	1,023	1915	773
1860	1,382	1892	998	1920	735
1865	1,282	1900	910	1925	688
1870	1,300	1905	912	1930	652
1875	1,256	1910	803	1940	685
1880	1,271				

Sources: U.S. Census and New York State Census population schedules.

within New York: 90.07 percent for Otsego and 95.63 percent for Schoharie, as compared with a statewide average of 68.14 percent. This preponderance of New York natives suggests that very little serious in-migration was occurring, as does the fact that the percentage of Otsego's inhabitants who had been born in the county itself was fully two-thirds, 67.39 percent, the seventh highest in the state, whose average as a whole was 53.33 percent. Similarly, Otsego showed a very low percentage of foreign-born inhabitants, a mere 5.34 percent, as opposed to New York State's substantial average of 25.45 percent.[3] Such homogeneity was owing to the basically unpromising local economy, which could not support the excess indigenous population, let alone those who might enter Otsego from elsewhere. In fact, few people entered the county from outside; many from inside left.

What this paring away meant for a craftsman like James Holmes is not easily discovered. Certainly one would not expect Westford to be an active center for carpentry, especially house-building. Holmes's own brisk pace in 1869 may seem to belie this observation, but in fact he must have been the only busy builder in the township that year. A brief notice in the Schenevus paper in August, submitted by a Westford resident, sketched the building scene for that year: "Five new dwellings are being erected in our village the present season," the note ran, "and some [others] are undergoing repairs."[4] Since Holmes himself was responsible for the erection of three of the new houses in Westford that season,[5] and also was responsible for some of the repairs, not much would have been left over for his competitors.

In any case, not many competitors were left. The departures of Milton Wright and his son Will for Schenevus and of Jud Winans for Oneonta, which took place in 1869 and 1870, offer simply the most particular evidence of a general exodus of craftsmen from Westford during Holmes's career. These three men evidently were responding to the new opportunities for work in Schenevus, a station on the railroad, and in Oneonta, where (as noted earlier) the location of the A & S Railroad yard and shops was to be a great stimulus to local construction. A comparison of the numbers of carpenters reported for different townships in Otsego County at midcentury will suggest how well the movement of the Wrights and of Winans reflected a wider pattern. Table 2 shows the relative gains of several nearby townships and two (Oneonta and Otego) at the opposite end of the county. Aside from Middlefield and Decatur, all the other townships owed much of their growth in the period from 1860 to 1870 to the coming of the railroad; even Decatur, which was not on the rail line but which was considerably closer to it than were Westford and Middlefield, may have owed its doubling of carpenters in these years to the same influence. As the average number of inhabitants per carpenter in these seven townships first rose and then dramatically fell in this twenty-year period, the ratio in Westford continued to increase, with Middlefield showing essentially the same pattern despite a slight improvement in 1870. In that last year, only these two townships in the southeastern corner of the county had ratios above 1:100 (the national average in that year was 1:113, and New York State's, 1:82). In the other five townships, between 1850 and 1870, the number of individuals per carpenter fell by a minimum of 19.1 percent (Worcester) and a maximum of 74.1 percent (Decatur), for an average decline of 51.1 percent. This means that the overall average decline for the seven townships as a whole in this twenty-year period, a drop from 1:116 to 1:78 for a decrease of 32.8 percent, was significantly affected by the increases for Middlefield (33.3 percent) and for Westford (21.0 percent).

Any carpenter who remained at work in those two townships, to be

Table 2. Total Number of Carpenters (C) in Selected Townships, Otsego County, Compared with Total Population (P), 1850–1870

Year	D	Myd	Mdfd	On	Ot	Wfd	Wor	Totals
C, 1850	3	22	29	20	12	12	17	115
P, 1850	927	2,152	3,131	1,902	1,792	1,423	2,047	13,374
C:P	1:309	1:98	1:108	1:95	1:149	1:119	1:120	1:116
C, 1860	5	19	19	13	15	10	16	97
P, 1860	902	2,225	2,825	2,158	1,957	1,382	2,134	13,583
C:P	1:180	1:117	1:149	1:166	1:130	1:138	1:133	1:140
C, 1870	10	45	20	44	32	9	24	184
P, 1870	802	2,402	2,876	2,568	2,052	1,300	2,327	14,327
C:P	1:80	1:53	1:144	1:58	1:64	1:144	1:97	1:78

Township key: D = Decatur, Myd = Maryland, Mdfd = Middlefield, On = Oneonta, Ot = Otego, Wfd = Westford, Wor = Worcester.

Source: U.S. Census population schedules.

sure, had less competition. But the number of jobs also was declining. This number is hard to measure, although when one compares Westford township with the county as a whole, it is clear that large jobs of the sort Holmes secured in 1869 were indeed few in Westford. Such a comparison can be made on the basis of total dwellings in any given area, figures for which are readily available in the state census for 1855, 1865, and 1875. Table 3 displays the data only for the townships used in table 2, along with countywide totals. Here the most striking change is that recorded for Oneonta between 1865 and 1875. As noted in chapter 1, that single township was responsible for 423 of the increase of 1,014 in reported dwellings in the whole county for that decade, or 41.6 percent of the overall change.

It would be wrong, of course, to assume that all positive or negative changes in these numbers between decades reflect actual changes in the housing stock in an area. For one thing, the instructions issued for these particular censuses defined a "dwelling house" as "a separate *inhabited* tenement" (my italics), so that any vacant structure ought not to have been counted.[6] An increase in the number of *families* in an area

Table 3. Numbers of Dwellings in Selected Townships of Otsego County, with Countywide Total

Year	D	Myd	Mdfd	On	Ot	Wfd	Wor	County Total
1855	166	398	551	407	388	275	387	9,287
1865	173	421	533	460	393	290	434	9,999
1875	167	427	565	883	453	314	532	11,013

Township key: D = Decatur, Myd = Maryland, Mdfd = Middlefield, On = Oneonta, Ot = Otego, Wfd = Westford, Wor = Worcester.

Source: New York State Census population schedules.

might cause previously unoccupied *dwellings* to be counted in a subsequent census, much as an exodus of population might reduce the number of occupied dwellings even though (hypothetically, at least) a substantial number of the now fewer occupied dwellings in fact might have been erected since the previous census. But it is unlikely that such a possibility as the last will occur with much frequency, especially since decreases of population are usually associated with adverse economic conditions. With these cautions in mind, we thus may assume in most instances that there is some correlation, however difficult to specify, between increased dwelling numbers and building activity.

But this is slippery ground. While Westford was losing population and carpenters, it was gaining dwellings and therefore showing a progressively smaller number of people per house. The case is more dramatically clear in Oneonta or Worcester, while the relative flatness of the Maryland township figures between 1865 and 1875, in light of the large increase in carpenters in the 1870 federal figures, is surprising. Other evidence from the period suggests that the figure 427 for 1875 may be in error. But this figure (rather than, say, its partial inverse, 472) clearly was used, as reported, in various calculations in the census tables, so that there would not seem to have been a printer's error. In any case, a whole neighborhood of new houses added to the village of Schenevus soon after the railroad arrived simply cannot have been reflected here.[7]

The county's average increase in reported dwellings across these periods was 7.6 percent from 1855 to 1865, 9.1 percent for the following decade, and 18.6 percent overall. Westford's figures (5.4, 8.3, and 14.2 percent), even though clearly well behind nearby Worcester's (12.1, 22.6, 37.5 percent), were not far below the county level. But Westford's "average" status here is tricky, for the countywide figure reflects the negative figures of several markedly declining townships in other parts of the county. The average change in these declining townships for the years from 1865 to 1875, −3.3 percent, reduces the county's average increase from 16.4 percent to 9.1 percent for that period. In reality, then, among those townships that increased, Westford experienced only about half the average increase. Its increase of 8.3 percent was considerably above the lowest average increases (.22, .86, 1.4, and 3.2 percent) but was even farther removed from the highest average increases (22.6, 26.6, 41.1, and 96.1 percent).[8]

It is clear from these figures that the building of five new houses in Westford village in the year 1869 alone must have marked a strong variation from the norm across this period. We learn again just how unusual Holmes's achievement that year was. If he had been equally active in each year from 1865 to 1875—building, shall we say, three-and-a-half houses each year (to allow for the carryover of the Winans job from 1868 to 1869)—he alone would have been responsible for adding thirty-five new houses to the township, when in fact an increase of only twenty-four dwellings was reported.[9]

Another means of contextualizing the Westford carpenter's situation is to add to such statistical information more descriptive evidence gathered from the local press. Especially in the smaller villages of the period, it was not uncommon for the local paper to report on many construction jobs as they were initiated, as work progressed, and as they were brought to conclusion. A somewhat broader conspectus of wide areas of the county was provided at the start of both 1868 and 1869 when the editor of the *Cooperstown Freeman's Journal* sent around systematic inquiries regarding building activity during the previous year.[10] Although the answers he received each year were scattered, they do provide a concrete picture of the building trades in the period just before Holmes kept his diary. During 1867, for the seven villages for which information was supplied, a total of about $200,000 worth of new work and substantial repairs was reported, ranging from a low of $2,000 for the new hamlet of East Unadilla to a high of $81,000 for Otego. Since the newspaper omitted any calculations for the county seat and apparently received no information for the other fifty or so villages in the county, these data are obviously sparse. Still, the particulars for a couple of villages (Unadilla and Otego) are rather full, and the information for three villages near Westford in the southeastern part of the county (Schenevus, Worcester, and East Worcester) is revealing. Like most of the smaller settlements in Otsego County, Westford itself is not reported in either of the annual summaries.

The type of work done in the reporting villages varied considerably. In Unadilla during 1867, much of the work was for updating or converting old structures rather than for erecting new ones. Of the thirteen structures on which major work was performed, only two—a new tailor shop and a small barn—were completely new, and these accounted for only a very small part of the total money spent in the village that year. One other small job, which entailed putting an addition on an existing store, was also new work, but it, too, did not cost very much. Of the six dwellings worked on during 1867, two were large frame houses that were "made over" in a modern style at the fairly high cost of $2,500 each. Four other houses that were repaired for amounts ranging up to $1,500 evidently required major alterations. Another building, previously used as a store, was converted to a house for $500, while a workshop was made into a store for just $300. In addition, the note on Unadilla adds (with a bit of boosterism), a good deal of sprucing up had been done across the village—much painting and laying of walks—besides miscellaneous repairs totaling about half as much as the amount spent ($12,500) on the major jobs.

In Otego during the same period, if we can believe the report, roughly four times as much was spent, a great deal of it on new construction. The Unadilla reporter in the following year, glancing back over the previous article, claimed that the Otego figures for 1867 in fact were a summary of work done over several previous seasons. Information available elsewhere regarding some of the jobs reported for Otego during 1867 does suggest that the figures for that year had been some-

35. Birdsall houses, 1878 view, Otego, New York

what padded. But it still is true that Otego was more active, at least for its size and perhaps in absolute terms as well, than Unadilla. Some reasons for this disparity will be discussed below. It should be recalled for now that in the township of Otego, the number of carpenters, reported as twelve in 1850 and fifteen in 1860, surged to thirty-two in 1870: the last figure may confirm, if not the specific level of activity reported for 1867 in the *Freeman's Journal*, at least the general increase in Otego building activity in the immediate postwar years. So, too, the increase in reported dwellings in Otego township between 1865 and 1875 (the figure jumped from 393 to 453) may offer further confirmation here.

The report for Otego's 1867 activity paints a bold picture of growth. Nineteen new houses were built, three at the cost of $1,000 each, fourteen at $2,000, and two at $3,000. In addition, an Albany lumber merchant who was a village native, William Birdsall, put $15,000 into erecting two new dwellings, plus barns and wagon houses, on the old family farm he owned just west of the village proper. Here he was soon to make his own home (fig. 35). While this particular set of buildings represented a rather atypical kind of activity in the area—an early case of gentrification, as it were—the other work reported for Otego during 1867 seems to have been more usual. Three new commercial structures (a warehouse, a modest shop, and a store) were erected at a total cost of $4,000, along with a multiroom, graded school that was just being completed early in 1868, which cost another $4,000 by itself. Repairs on houses and commercial structures, all of them itemized in the report, amounted to a full $18,000, almost as much as was reported for all work in Unadilla the same year. Moreover, another $7,000 was spent in Otego repairing four church buildings. The grand total for the village thus ran up to $81,000.

Even if, as the Unadilla reporter claimed in 1869, this figure did cover "all the building done for the past three or four years" in Otego,[11] it remains evident that Otego was at least as active as, if not more active than, its larger neighbor. Hence it should be clear that it is difficult to generalize about building activity even in so constricted a geographical region, across which larger economic forces tended to act rather equally. Only seven or eight miles apart on the old wagon route and the new railroad that roughly followed it, the villages of Unadilla and Otego occupied similar sites on the Susquehanna River, where waterpower was available, if not abundant, and where the hilly terrain of the surrounding countryside was broken by both the river valley and the intervales of tributaries coming in from the north. Unadilla was, however, both older and larger, having been settled before the Revolution and incorporated in 1827, quite early for this area. If anything, it might have been expected to have the edge over Otego in postwar development. A trend in population in the recent past, though, seems to have favored Otego. Between 1865 and 1870, the township of Unadilla lost 130 people, while the township of Otego gained 169, a contrast that is even more striking when the beginning figures for the two in 1865 (2,685 for Unadilla and 1,883 for Otego) are taken in account. The numerical gain for Otego over Unadilla township in those years was 299, but when it is corrected for the relative standings of the two townships in 1865, it was equal to a gain of 440 in Unadilla.

Figures for the villages proper are harder to come by for these years. An 1840 estimate puts Unadilla at 365 and Otego at 200. In 1860, the two were put at 795 and 311, respectively.[12] At that point, Unadilla's growth rate was considerably higher than Otego's (117.8 percent as compared with 55.5 percent), but soon the balance was to shift. The newspaper report on building activity in 1867 states that Otego had increased by half in the years since the war, although it gives no figures.[13] Figures supplied in the federal census (from 1870 on for Unadilla, from 1900 on for Otego) show that this claim had, over a longer period of time following, a certain continuing validity. The population of Unadilla village rose from 875 in 1870 (a gain of just 80 people over 1860) to highs of 1,157 and 1,172 in 1890 and 1900, after which it fell to 1,009 in 1910. Its growth rate from 1860 to 1900 thus was 47.4 percent, but from 1860 to 1910 it was only 26.9 percent. Otego, on the other hand, rose from 331 in 1860 to 658 in 1900 and then to 676 in 1910, so that its rates of growth were 98.7 percent and 104.2 percent, respectively, for the same intervals. More interestingly, the absolute gains in Otego over the full fifty-year period (345) were 61.2 percent greater than those in Unadilla (214) in the same era. Between 1860 and 1910, in other words, Otego village more than doubled in size, whereas Unadilla added only about 25 percent to its original population. And even though Unadilla remained larger than Otego at the period's end, Otego's gain in absolute terms was more than one-and-one-half times Unadilla's. Interestingly, over the same five decades, the population of the townships in which these villages were located dropped at roughly the same rates: Unadilla's dropped from 2,702 (1860) to 2,601 (1900) and 2,376 (1910) and Otego's from 1,957 (1860) to 1,817 (1900) and 1,699 (1910), for rates of decrease of 3.7 percent (Unadilla) and 7.2 percent (Otego) in the shorter period and 12.1 percent (Unadilla) and 13.2 percent (Otego) in the longer.

There was a contemporary perception of some such differential between the two rival villages.[14] D. Hamilton Hurd's 1878 *History of Otsego County* portrays Unadilla as a rather quiet place more notable for small-scale improvements on its appearance than for any dynamic growth. The improvements included its "miles of stone walks," a novelty in rural villages at the time, as well as its "fine" streets, which were in good repair and were "beautifully shaded" by trees lining the walks on either side. Balancing these public improvements were private ones that made "many of the residences and grounds in Unadilla among the

36. Otego, New York, view along Main Street, ca. 1910

finest in this section." For Otego, on the other hand, Hurd's *History* spared the flowery diction used for Unadilla ("art has vied with nature," runs one typical orotundity perpetrated there) and came right to the point: this was, the *History* asserted, "one of the flourishing towns [i.e., villages] of the county." "Flourishing" was the usual code word then for smart economic activity, and it does seem to have been an appropriate term for describing Otego at this point in its career (fig. 36).[15]

For one thing, the preponderance of new work over repairs in the *Freeman's Journal* report on Otego supports such a description. A glance at the 1868 Otsego County atlas (fig. 37) will reveal with some prominence what was happening to the landscape of the village right after the war. Averill, Follett, and Willow streets all were laid out in 1867, the newspaper report indicates; in 1868, apparently, the completely blank stretches of Newland and Saunders streets were added. While much of the land along these five streets remained vacant at the time the atlas was published, it nonetheless is significant that the village had basically doubled its total length of streets in a very short period of time. And it was along these new streets, especially those between the

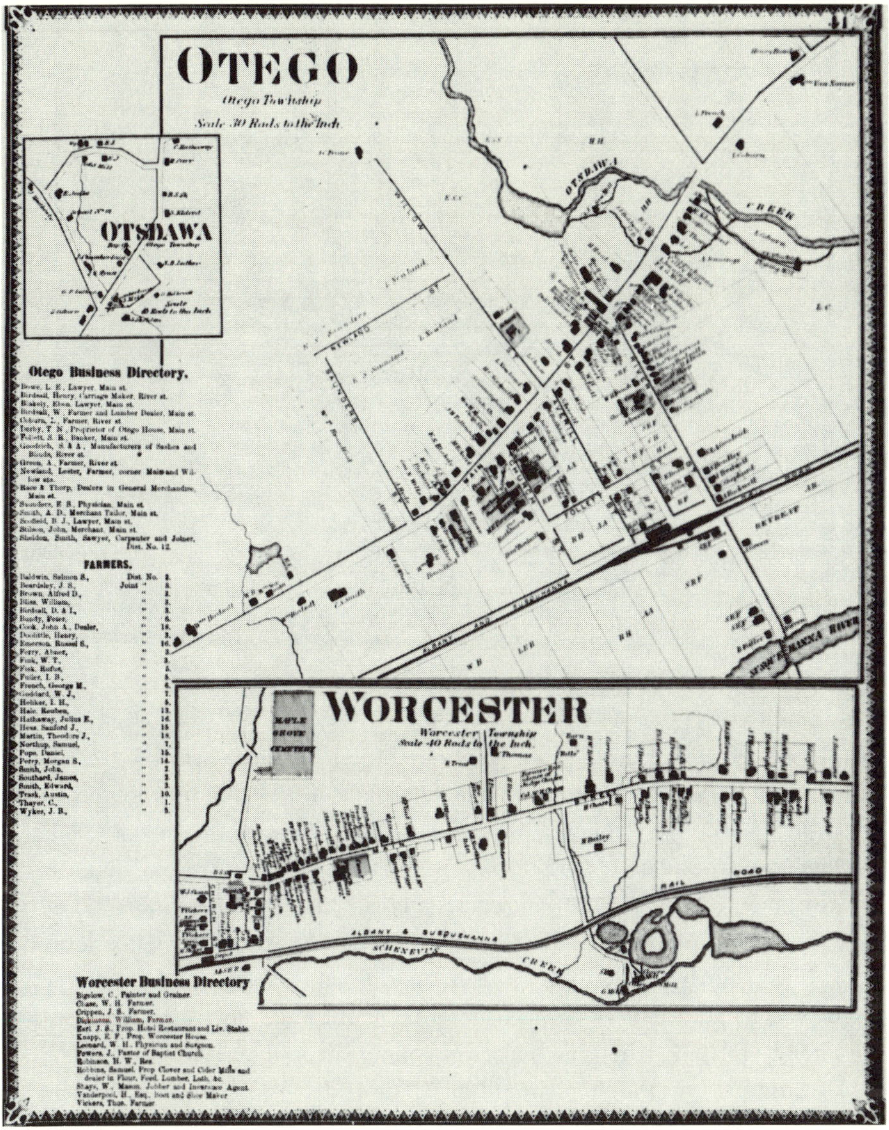

37. F. W. Beers, maps of Otego and Worcester villages, 1868

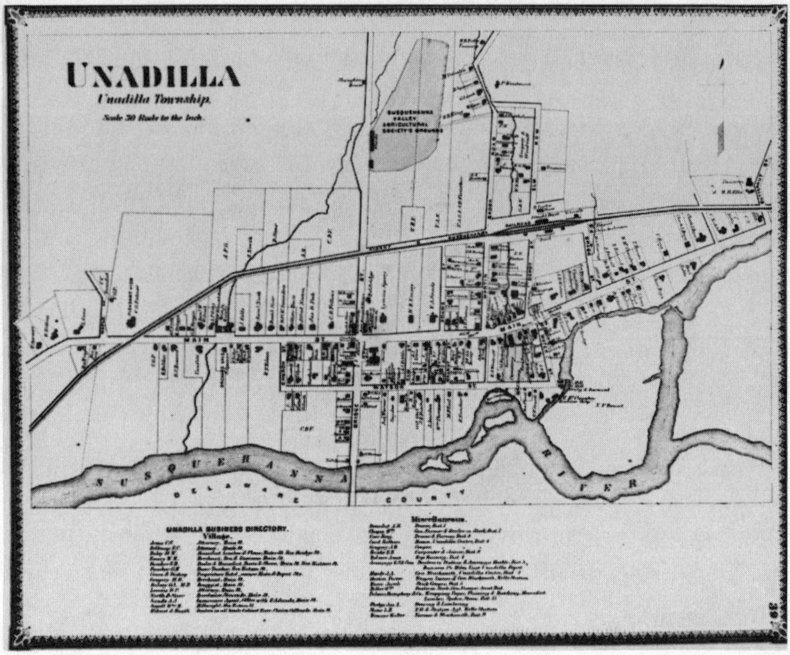

38. F. W. Beers, map of Unadilla village, 1868

old village center on the stage route and the new depot near the tracks at the village's southern flank, that the bulk of new building centered in the period. On Follett Street, for instance, five of the new $2,000 houses were erected, perhaps on speculation, since no owners' names are given in the newspaper report. Plans for new work in 1868, the report also stated, had been developed for nearby Averill Street, where several new dwellings were to be erected. Church Street, with its share of vacant land toward the tracks on its south end, saw some activity already in 1867, when the houses of Daniel Birdsall and John Rathbun ("Rathburn" on the map) were built. The rest of the new construction in this quarter of the village was along River Street, where four houses and two commercial structures were put up. To the north, several new buildings, both dwellings and commercial structures, were erected on lots scattered along Main Street. Beyond this infill activity, the only major work north of Main Street in 1867 occurred on Willow Street, first named after J. C. Frone, whose house at its upper end was constructed that year. Not visible on the map are two houses built on the upper reaches of Terry Street, the unnamed road that joins Main Street just beyond Otsdawa Creek in the upper righthand portion of the map.

The Unadilla map from the same 1868 atlas (fig. 38) reveals a locality without such dramatic changes in process. One reason may be that the coming of the A & S Railroad introduced no focused spatial impulse toward expansion. The railroad came through the north part of the village, for one thing, where the land soon becomes precipitous. And the depot was located just above where the old village reached its narrowest point, so that little prime real estate was created by its erection. Certainly there were economic stimuli generated by the railroad, but the peculiar geography of Unadilla made the translation of those stimuli into spatial terms considerably more difficult than was true in Otego, where ample vacant and buildable land lay between the depot site and the village center. A similar effect can be noted, by way of comparison, in the case of Richmondville, a village just beyond the eastern Otsego boundary in Schoharie County. Here the A & S line came through on the hill face on the north side of the creek (fig. 39) across the valley from the village and significantly above its level. Over time, the presence of the railroad there had relatively little direct effect on the physical layout of the original settlement.

Certainly in 1867, the construction work reported for Unadilla had no visible relation to the coming of the train either. And the work reported for the village in the generally less useful 1869 *Freeman's Journal* article, although it was of greater monetary value and did involve a great deal of new construction, did not involve the laying out of any new streets in Unadilla, either near the depot or elsewhere. The ten new houses reported for 1868 show a considerable increase over the level of activity noted for 1867, but this new work seems to have been scattered around rather than concentrated. Besides, the average cost of these new houses was significantly lower than the average for the houses reported for Otego in 1867. And the total of new and repair work for Unadilla in 1868, $25,300, while it was a sizable increase over the $18,750 reported for 1867, still was far behind the $81,000 given in 1867 for Otego. Even if that latter amount were to be spread over five years, more than ample to take care of the criticisms of the Unadilla reporter in 1869, the average of $16,500 per year that this reduction would allow, if adjusted for the difference in size between the two vil-

39. Richmondville, New York, view from Frayer's Hill showing A & S Railroad line on hill opposite village, ca. 1905

lages, still would give Otego a substantial lead. It seems likely, furthermore, that five years is too generous a concession to the rival Unadilla reporter's suspicions. The smaller village quite simply was outbuilding its neighbor.

The effects of the railroad were also felt in villages closer to Westford. Activity in the three villages near Westford for which figures were supplied in the *Freeman's Journal* for 1867—Worcester, East Worcester, and Schenevus—resembled Otego's more than Unadilla's. The actual amount of growth in each of these villages was more modest. But its nature and location make the comparison with Otego quite fitting. There was a good deal of new construction, much of it was on new streets between the old village center and the depot to its south along the banks of Schenevus Creek, and all this growth entailed an investment proportionally (or even absolutely) higher than that in Unadilla. In East Worcester, a total of $36,400 was reportedly spent, only $5,400 of it on repairs. Two new streets were opened up between the wagon route and the train depot, as visible on the 1868 map (fig. 40). The

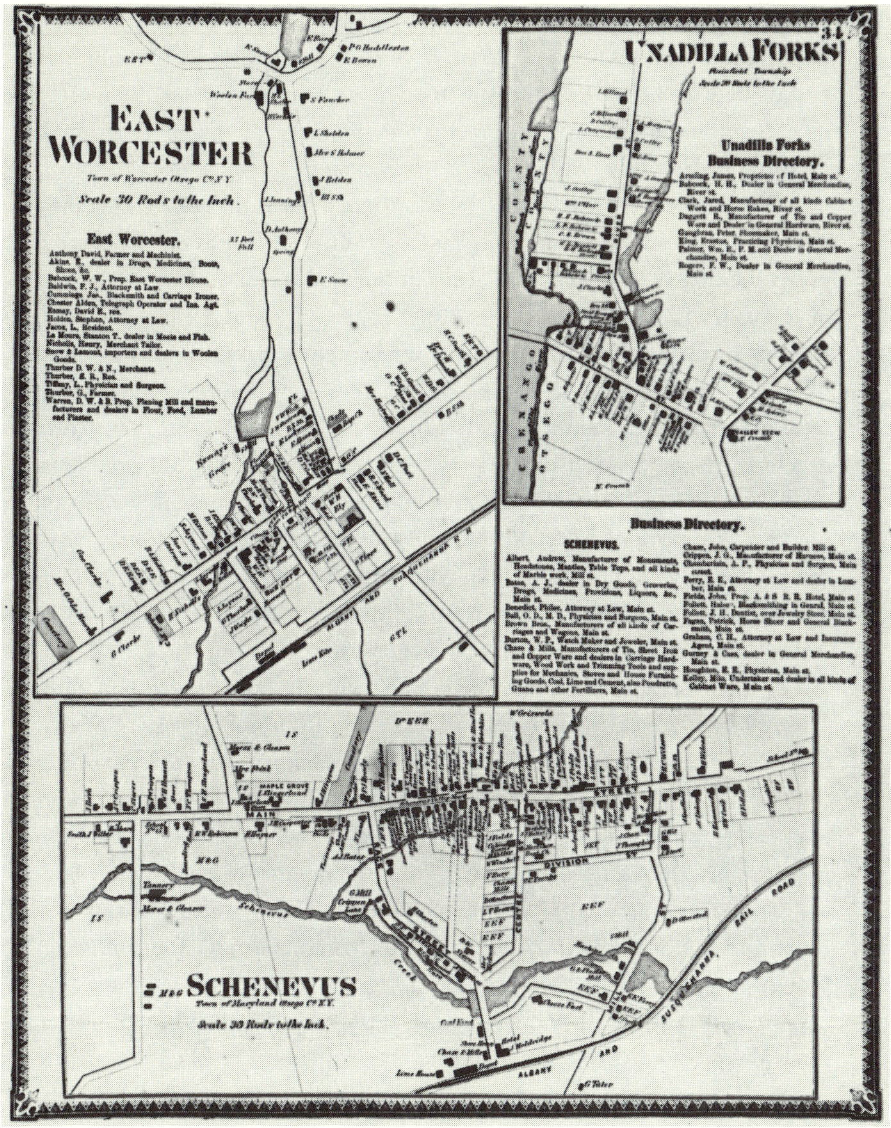

40. F. W. Beers, map of East Worcester and Schenevus villages, 1868

extent of the expansion in East Worcester, a village noticeably smaller in population than Otego (in 1840, they were estimated at 150 and 200 and in 1872 at 393 and 600, respectively),[16] was more modest than Otego's. But the amount invested in East Worcester was nearly twice that invested in Unadilla, which was much larger still.

In Worcester, a few miles to the west, the activity in 1867 was somewhat lower in dollar value and was fairly modest in effect, especially given the fact that Worcester earlier in the century had been the larger of the two villages.[17] But the $30,000 reportedly spent there in 1867 was no mean amount, particularly by comparison, again, with Unadilla in both 1867 and 1868.[18] Here, fourteen new dwellings, costing a total of $26,000 and running individually from $1,200 to $3,000, were erected. Although most of these were spread around the old precincts of the village, and only three were built on the single new passageway opened up between Main Street and the depot, the report promised more expansive activity the following year. But the 1869 report in fact showed a smaller dollar value for work finished in 1868 (about $18,000), as well as a smaller number of new houses, just six.

The modest nature of this work was owing in part to the fact that Worcester had a peculiar layout that tended to inhibit the kind of expansion seen in some of the other villages along the A & S route (fig. 41). It was the most extreme case of a strip village in the lower part of the county—the 1868 report called it "almost a continuous village of more than a mile in extent"—and its modest center was merely a slight thickening of this line near one end. The closeness of the rail route and the depot to this village center, combined with the confining topography of that immediate vicinity, made it more difficult for the expansion to square off the strung-out settlement, as it did in Otego or East Worcester by sending longish perpendicular streets toward the tracks. The 1868 atlas map (fig. 37) reveals both the original elongation of Worcester and the problems posed by the location of the A & S line there. There was no natural "new" neighborhood adjacent to the railroad.

This constraint on Worcester's expansion did not keep the village from seeking the kind of urban finish that the larger settlements of the

41. Worcester, New York, view looking north past A & S Railroad tracks, ca. 1905

county hungered for in the postwar years. On the side of the hill north of Main Street, just after the war, Worcester had begun developing Maple Grove, a new burial ground in the current parklike style so well exemplified by the rolling expanses of the large rural cemetery outside the state's capital city, Albany. And, like Unadilla, the village was spending funds on more utilitarian repairs, public improvements, and even quasi-urban amenities—the kind of thing a village of Westford's size never could hope to achieve. The effect of such changes in Worcester was not lost on outside observers. Andrew Boyd, in the first issue of his *Albany and Susquehanna Railroad Directory* (1868–1869), a business compendium that signaled the economic importance of the new rail route, proclaimed boldly:

> The village of Worcester, which [heretofore] was found nestled between the hills in the Schenevus Valley, has shook [*sic*] off her Rip Van Winkle dreaming, and waked [*sic*] up to the realities of life, which is made visible in the new and attractive features of the place. Streets have been straightened and graded, beautiful shade

trees set out at proper distances apart on each side of the road, new streets laid out and opened, and fine stores, dwellings and mechanic shops have been erected and finished according to the modern style of architecture. Numbers more are being built and others [are] in contemplation of erection, older dwellings have been repaired, remodeled and newly painted, rendering the place very beautiful and attractive to the pedestrian and stranger.[19]

BUILDING ACTIVITY IN SCHENEVUS, 1868 – 1870

For Westford's residents, the new scene in Worcester ought to have been readily familiar: Worcester was the closest large village, after all, and its depot provided the nearest access point for the railroad. But as James Holmes's 1869 diary suggests, Worcester was not as easy for Westford residents to reach as was the actually more distant village of Schenevus. Positioned on the upper reaches of Elk Creek, Westford was separated from Worcester by high ridges that made the wagon route there—the road on which Holmes built his new house in 1869—a rather difficult one. It was easier to use the relatively flat route that went south out of the village, following the widening valley of Elk Creek, toward Schenevus eight miles away. Hence, the economic and social ties between Westford and Schenevus had always been fairly strong. And as Schenevus responded vigorously to the coming of the railroad, those ties became even stronger.

In Schenevus itself, as in Otego at the opposite end of the county, the impact of the railroad was felt in strongly physical as well as economic terms. The growth reported for Schenevus for 1867 in the 1868 *Freeman's Journal* story, to be sure, seems rather modest at first, even by comparison with that noted for Worcester or East Worcester, since only about $22,000 was spent. But much was accomplished for this sum. It covered the cost of erecting eight new unspecified buildings, as well as a plaster mill and a large cheese factory.[20] Two new streets—Centre Street, whose lots already had been divided by 1868, and the sketchier Division Street—had been laid out in the village to link the old center with the depot, which here lay across the creek (fig. 40). Soon the land south of the second of these streets, owned by Elijah E. Ferry—a lawyer and judge who ran a sawmill and other operations and also owned the new plaster mill—would be developed as well. The result in this case, as in others, was that the old string of dwellings along a single street, thickened slightly by a cluster of businesses near its center, was roughly squared by the coming of the rails. No longer was the attraction of waterpower the only reason to penetrate the adjacent land between the modest rise where the village stood, relatively safe from the flooding waters of Schenevus Creek, and the bottomland bordering the creek. Soon, with the presence of steam engines, that older incentive in fact would pass. But the train depot would anchor the faith of the village at its southern edge for another sixty years or more.[21]

The liveliness of the village as it responded to this revival of its economic life was much noted at the time. Hurd, in his 1868 county history, commented on the special effect that the A & S had on Schenevus in particular: "The business of the village rapidly increased, and it soon became one of the leading towns [i.e., villages] on the line of the road."[22] Incorporated in 1870, before Worcester, East Worcester, or Otego, Schenevus grew from a population of 383 in 1860 to 549 a decade later. Estimates during the 1870s placed it even higher—700 in 1872 and 726 or 800 in 1878—and although the official count in 1880 was lower than these estimates, it still was a surprising 665. After that point, a slow decline set in; although Schenevus never rose to the absolute level of Unadilla, however, it roughly paced Otego.[23]

Like both those other villages, to be sure, Schenevus hardly showed the meteoric rise associated with the village of Oneonta, which increased from a population of 678 in 1860 to 1,061 in 1870 and then to 3,002 by 1880. By 1910, in fact, when most of the villages in Otsego County had been in serious decline for some time, Oneonta had just been incorporated as the county's first city. It had a population then of 9,491, equal to the combined populations of the townships (including, of course, the villages) of Unadilla, Otego, Maryland (i.e., Schenevus), and Worcester, with Westford and Decatur thrown in. Without the presence of so direct a benefit from the railroad as Oneonta had in the A & S shops, villages like Schenevus had little hope for such substantial urbanizing transformations. Even so, among villages of the middle rank, Schenevus prospered well enough for a time following the late 1860s.

One reason was that it was an ambitious village, with high hopes and much activity. As the war was ending, it saw the establishment of the Schenevus Valley Agricultural Society, an organization that reached out to embrace farmers and mechanics in the main valley villages, from Maryland and Chaseville and Milford on the west through Schenevus and Worcester to East Worcester in the opposite direction, plus the tributary hamlets of Middlefield to the west, Elk Creek, Decatur, and Westford to the north, and several townships in the upper part of adjoining Delaware County. It was the fifth annual SVAS fair that James Holmes attended in October of 1869, on what was one of his very few holidays from work. Such festivities could not compete with the annual Otsego County fair in Cooperstown,[24] one of the oldest in the nation; but they showed the intention of Schenevus to become a leader in its own right and to contribute as it could to the improvement of local agriculture and industry.[25] Similarly, the village sported a number of social organizations, most founded in this postwar period: a cornet band and a circulating library, as well as Masons, Odd Fellows, and Good Templars lodges and a Grand Army of the Republic post, a rather sizable array by local standards. Finally, it had not only the single newspaper that had been founded in 1864, the *Monitor*, but a second one begun in 1868, the *Schenevus Valley News*. At this time, Worcester, East Worcester, and Otego all were without even a single paper.[26]

The founding of this second paper, in fact, allows us to gauge with considerable detail the unfolding development of James Holmes's immediate world, since the *Valley News* paid careful (and obviously interested) attention to the transformation of Schenevus and its vicinity, including Westford, in these years. So closely did the *Valley News* scan the developing village scene that toward the end of August, in its first surviving issue, the paper quite typically devoted space to the "news" that the hardware merchant Samuel T. Chase of Chase & Mills—with whom Holmes had dealings in 1869—was putting up a new fence in front of his dwelling[27] and that Mrs. Caroline Cyphers was having an addition put on her house.[28] As these modest facts suggest, very little that arguably contributed to the increasing finish of the area could escape the paper's notice.

Over the course of the next couple years, the *Valley News* continued to monitor with similar attentiveness the changing face of the settlement. It allied itself from the beginning with the railroad, seeing it as the vehicle by which the depressed locality might tie itself, profitably of course, into the larger market economy that the State of New York so eminently typified. Hence in June 1869, it proclaimed that "Otsego County, long pent up in the wilderness, is fast emerging from her torpidity, and is destined to rank among the foremost counties in the State in producing supplies for the markets." In March 1869, just as Holmes in Westford was beginning his jobs for the new year, it proclaimed that "building operations will be very active in all railroad villages of the country [county?] the entire year"; "railroads," it went on, "[mean] business, growth and prosperity everywhere."[29] But the newspaper was not immediately alert to what was probably the biggest result of the railroad's arrival, a result that in Schenevus and the surrounding countryside was felt in agriculture. For it was farmers such as Joseph Marvin Holmes, more than craftsmen such as his younger brother James or the shopkeepers in Westford or Schenevus, who would profit almost without delay from their new ability to export dairy and other products rapidly to urban markets. In the single month of November 1869, the *Valley News* reported the following month, a half-million pounds of freight were sent from the Schenevus station alone, including more than a quarter-million pounds of butter (worth some $80,000 in prices then current), the greatest quantity of any single product shipped from the village then. Another hundred thousand pounds of vegetables and fruits, and ninety thousand pounds of hops (worth more than $150,000 that year) also came through that station, increasing the farmers' apparent gains from the new line. The rest of the freight, some fifty thousand pounds, was composed of miscellaneous items that obviously made up—with whatever industrial or commercial products were included—a very small portion of the whole.

Still, it was not just farmers who stood to gain from the coming of the rails, as the March 1869 comment in the *Valley News* about building activity suggested. As long ago as 1864, when the line had reached from Albany only as far as Schoharie, some thirty miles away, the *Schenevus Monitor* already had begun sounding this same theme: "The growth of the village has, necessarily, been slow, with nothing but common team roads, to and from market, to compete with thoroughfares, north and south of us;[30] but a new and gratifying [era] is dawning upon us; the iron horse is soon to tramp the iron way, and the 'good time is coming.' "[31] The writer's further point in this 1864 letter to the *Monitor* was one that came to guide the practice of the *Monitor*'s competitor once the train actually arrived. The "accelerated progress and improvement" of the village, he declared, would be "evidenced by the erection of more and better buildings."[32] It is not surprising that the *Valley News* chronicled the physical alterations in Schenevus with great care. Between its founding in the late summer of 1868 and the end of the following calendar year, it reported on the building of at least fifteen new houses in Schenevus. Four more, including that of the paper's publisher, Hiram T. Oatman, were mentioned once as being planned, but no follow-up stories were found. The paper also described the conversion of two shops into houses and detailed the improvements carried out on another eighteen houses, ranging from painting and papering to the addition of verandahs or substantial wings. Finally, it even chronicled the erection of some half-dozen outbuildings—barns, sheds, wagon houses, and the like.

The spacing out of this information through the pages of some forty issues of the *Valley News* allows the reconstruction in a rather contingent way of the village's progress over the seventeen months in question. Such a reconstruction usefully parallels the work of Holmes in Westford.[33] Since the paper began publication as the 1868 season had passed its peak, its record for that year is rather spotty. Still, the August 25 issue, the first surviving one, did comment on Samuel T. Chase's fence and Mrs. Caroline Cyphers's addition, as well as the laying of the cornerstone for the Baptist church. The issue of the following week reported that the dressmaker Mrs. Mary A. Kelley was having a cellar hole dug and was otherwise preparing to convert her commercial building into a dwelling; it added as well that the new church's frame had been erected on the previous Thursday. One week later still, the church had been enclosed and shingled, and on the day the paper came out a new house frame was to be raised on Water Street, near the depot. In mid-September, the paper noted that carpenter John Chase's new house on East Street was nearly completed. A week later, on September 22, it reported that the move to build a graded school in Schenevus—this would be another mark of the village's modernity—had culminated in the purchase of a lot on Centre Street and in the election of three trustees, one of them the Westford native (and carpenter's son) Alden Chester. (As it turned out, the vote on this graded school was later reversed; but such a school was in operation a decade later when Hurd's *History* was published.) In the same issue, the paper reported that Oscar P. Toombs, a marble-cutter and monument-dealer in the village, had begun a new house near the Baptist church that would "prove to be a credit and ornament to the village." Mention also was made of the decision recently reached by the hardware merchants, Chase & Mills, to erect a sixteen-by-twenty-two-foot addition to their store in order to handle the heavy business they were experiencing. Finally, the last issue of the month of September noted that John Holdridge, proprietor of the new hotel near the railroad depot, was erecting a barn and some sheds near that property.

By October, when not much old work was continuing, the paper's publisher, Hiram T. Oatman, announced that he had let a contract to the carpenter George Witt to build him a new house on Race Street. The structure, to be of the size and style of wagonmaker Washington I. Waterman's house (the one raised on nearby Water Street early the previous month), was to be completed by January 15.[34] In November the paper reported that the wealthy farmer Hiram Banner was fixing up his house and erecting a large shed and wagon house, while in December there was word that another lot on Race Street had been sold and a house would be built on it in the coming season. Another house likewise was being planned for a lot on Main Street, while the Baptist church, finished up in the autumn, was to be dedicated at the very end of December.[35]

As the new year began, the *Valley News*, which in November had reported the imminent completion of the railroad with a sigh of relief ("Good-by, stages"), announced that the line was to open formally all the way to Binghamton on January 12. And more news of upcoming projects was emerging, even though the weather right then was bad, so that little actual progress could be made: New Year's Day had been awful in Albany, where Holmes then was, and in Schenevus it was "the most boisterous and blustering day of the season" to date. As for the new jobs, marble-worker Andrew Albert had contracted for John Chase to build him a new house on Mill Street. Already, the millwork establishment of Lane & Hotchkin was busy making a large quantity of sash, blinds, and doors for the many projects expected in 1869.[36] On January 12, the marble-cutter Oscar P. Toombs moved into his new house on Centre Street; it was "the nicest residence" on that street, the paper proclaimed, and "an ornament to our village." Then, except for the note on February 9 that the peddler John Fink had bought a lot and planned to build a house just south of the rail depot, there was a lull in building news. The weather turned mild and then stormy again, only to repeat the pattern in mid-February and early March, when it proved "unseasonably cold."

By the middle of March, projects were again being talked of. Cooper Peter J. Brady had bought A. R. McCourtie's shoe shop next to the Methodist parsonage and across from the cemetery on Main Street, plus an adjoining vacant lot, and was planning to convert the structure into a house. Across the way, a Mr. Squires of the village of Maryland had bought another vacant lot and was planning to erect a house there in the spring. Timothy Parsons, partner in the dry-goods, drug, chemical, and grocery firm of Parsons & Tompkins, had purchased hotelkeeper John Fields's vacant lot on Centre Street for the purpose of building a new house. And carpenter William E. Wright and his wife Sarah Holmes Wright were planning to erect their new dwelling south of Toombs's house on Centre Street. There were so many prospective jobs that it was no wonder Lane & Hotchkin was still busy turning out millwork items for the season ahead (fig. 42). So busy were they that late in March they made arrangements for an Albany glassworks to ship them ready-glazed sash, which the Schenevus house-, sign-, and carriage-painter William J. Merrihew would prime and paint on order. Although many local residents had been struck recently by "Western Fever" (the paper was continually reporting the destinations of Otsego

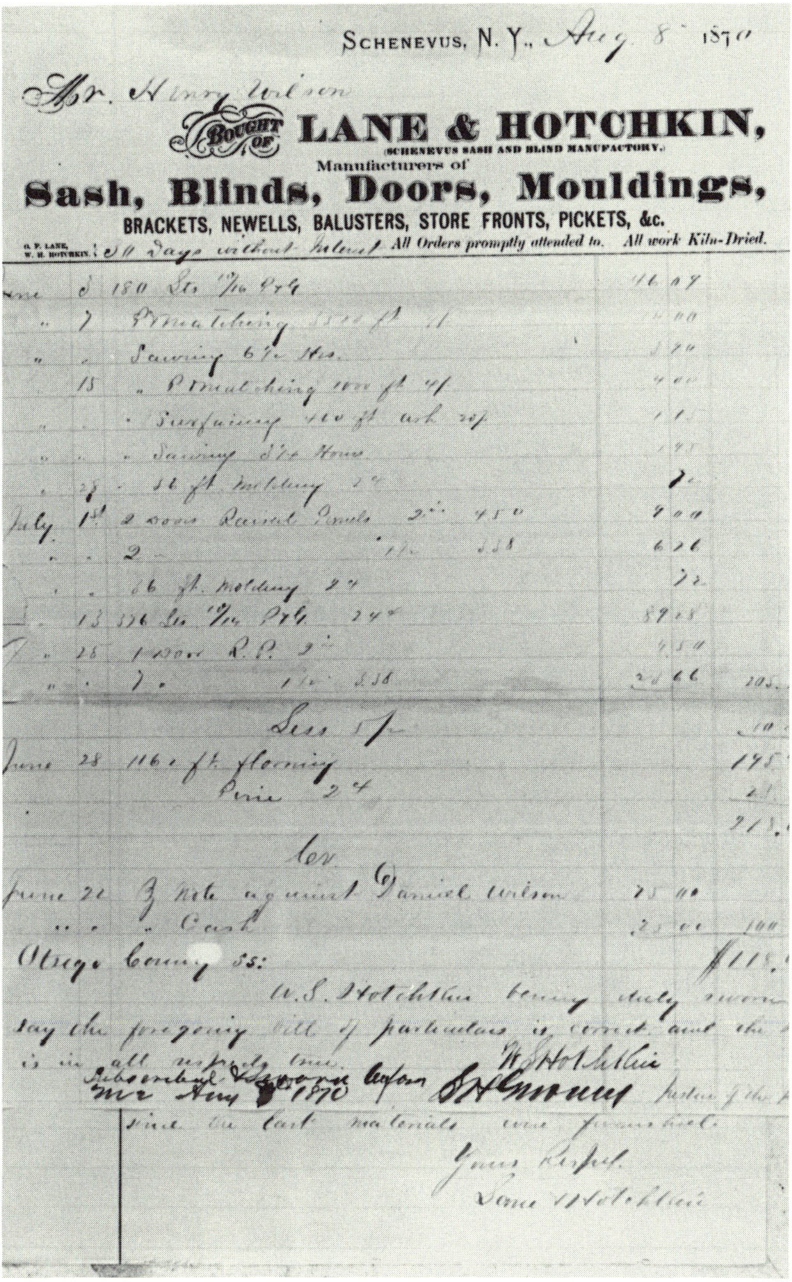
SCHENEVUS, N. Y., Aug. 8 1870

Mr. Henry Wilson

BOUGHT OF LANE & HOTCHKIN,

(SCHENEVUS SASH AND BLIND MANUFACTORY.)

Manufacturers of

Sash, Blinds, Doors, Mouldings,

BRACKETS, NEWELLS, BALUSTERS, STORE FRONTS, PICKETS, &c.

G. F. LANE, W. H. HOTCHKIN. 30 Days without Interest All Orders promptly attended to. All work Kiln-Dried.

Otsego County ss:

Yours Respect.

Lane & Hotchkin

42. *Lane & Hotchkin billhead, 1870*

residents who were headed for Illinois or Iowa or Kansas), clearly a substantial number of people were willing to invest their capital and their lives in Schenevus.

At the end of March, the paper reported that Dr. Ogilvie D. Ball had bought a property he planned to repair and paint. Word also was abroad that the "old ark" (obviously a notorious local structure) was being refurbished for use as a school to be opened in April by Celida Wright Hotchkin. Milton and Louisa Wright's daughter, Celida, just recently had married one of the partners in the prosperous sash-and-blind factory; her sister, Helen Wright Lane, had just married the other one. The last two issues of the *Valley News* in March added that the tailoress Mrs. A. H. Rathbone and the carpenter George W. Nellis were planning to have new houses built on Centre Street that summer. The first April issue noted the awful weather ("Snow partly disappeared—mud knee deep—rainy weather—just enough of each to make it very disagreeable") and then added that the stone blacksmith shop had been reopened by a Cherry Valley firm and that the wealthy merchant John T. Thompson was proposing to open a commission business later in the year, buying fresh produce from local farmers for cash.

By mid-April, the weather was springlike, and work was beginning in earnest. About two weeks before James Holmes began a similar task at his homesite in Westford, the men working for Mr. Squires were drawing stone and digging a cellar for his new house near the cemetery and the Methodist church. A velocipede rink had been opened recently on the upper level of a village shop, presumably necessitating some alterations to that structure ("How to ride the velocipede," the paper had advised when first reporting on the device's arrival in March—"Straddle the saddle, paddle, skedaddle"). Meanwhile, grocer Alexander J. Bates had begun adding a piazza to his house and painting the whole structure, and Dr. Ball was erecting a barn and painting and papering his recently purchased home. One week later, the *Valley News* could report that attorney Charles H. Graham had begun building a new office for his insurance business near his residence, which was located along the middle stretch of Main Street. Marble-worker Andrew Albert's cellar had been dug and the foundation was soon to be laid; a warm rain had been followed by sunny weather perfect for that task.

Meanwhile, Jacob Multer, publisher of the paper's rival, the *Monitor*, was raising the roof on his house and otherwise updating it, and William H. Bennett, a village tailor, was having such major repairs done to his shop on the lower (that is, southwestern) end of Main Street that he temporarily was doing business out of the old Boyce store. Elijah E. Ferry's sawmill along the creek was understandably busy at this time; it had turned out a great deal of lumber already that year, but it still had some logs on hand and could take orders for custom work.

As May began, farmer Barlow Patrick was repairing his house on the corner of Centre and Main streets, wheelwright Patrick Fagan and carriagemaker Washington I. Waterman were just beginning to run the blacksmith shop they had taken over from the Cherry Valley firm, and Fagan was contemplating painting his house and refencing his yard. Milo Kelley was having a new foundation prepared for his cabinet warehouse, which was to be moved to a new downtown location, while his brother and financial partner, the liveryman S. Lewis Kelley, had just finished adding blinds to his house and fixing its walk. Nelson Lane, partner in Schenevus Mills, likewise was planning to erect a new fence near his house at the corner of Main and Race streets.

Later in the month, the paper reported on the very high level of construction activity going forward in Cooperstown that spring. There, the new hotel, the Clark mansion, a sash-and-blind factory, and a number of "cottages" were giving employment to a hundred carpenters, masons, stone-cutters, painters, and laborers. But Schenevus, though its projects were more modest in scale, hardly was a sleepy backwater. Instead of moving Milo's cabinet warehouse, the paper noted, the Kelley brothers had decided to erect near the village's center (across from the Post Office) an entirely new structure. This would be a two-story thirty-by-sixty-foot brick building with a full basement, the upper story to function as a hall for exhibitions, lectures, and other public events. The paper strongly supported this aspect of the project: it was, after all, just the sort of grand idea that the *Valley News* would see as evidence of the proper public spirit in Schenevus.

But the paper also kept reporting on the smallest changes in the built environment, as if not just grand notions but also tiny adjustments might signify the community's new vigor. It thus noted that Egbert

Crippen, Lane's partner in Schenevus Mills, was putting up a new picket fence at his house on Race Street. And it added that John Field, the proprietor of the Susquehanna House, had recently "fitted up" that hotel and—surely so that he could compete with Holdridge's new hotel along the tracks—was planning to run a carriage to the depot this season. Soon the paper could report that Patrick Fagan indeed had fenced his house anew and was painting the structure and that lawyer Charles H. Graham was painting his as well. John T. Thompson had bought the vacant lot next to his store on Main Street, it noted, with plans to erect a bank building there, the second floor to house rental offices. And by late May it reported that the frame for the Squires house near the cemetery was up, that Timothy Parsons's cellar walls were being laid up at what the paper now called "the best location" on Centre Street, that Toombs was adding a picket fence around his new place farther up the way, and that talk was going around about the desirability of converting the cheese factory to a starch-making plant.

By the start of June, William H. Bennett's tailor shop had been taken down to its foundation and rebuilt to two stories, and a large wing had been added as a dwelling. At the same time, Mrs. Caroline Cyphers was having the steep yard behind and in front of her house graded, perhaps with plans to fence the whole. And the Kelley brothers' project was now set to begin the first of July, or as soon as the brick could be made—evidently by John Hubbard, who ran the village brick kiln south of the depot. Soon the paper was noting that boardinghouse keeper Frank Tompkins's new dwelling on Race Street had been raised and that Timothy Parsons was to raise his own the present week. The yard and grounds at the new Baptist church had been graded and a walk projected from that point up toward Main Street. William H. Bennett's building plans for the year soon were expanding, the paper reported on June 8. He already had part of the foundation laid for a house he was to erect on a parcel just outside the village at the upper (or northeastern) end. Around town, people were setting out evergreens and maples in their yards and along the streets; the paper welcomed this genteel touch and urged that the example be imitated widely. At the midpoint of June (when James Holmes and his helpers were moving inside at his own job) the *Valley News* noted that the sash-

and-blind factory (which recently had begun to manufacture the "Wash Room Economist," a washing machine) was in need of two or three good workmen. It also took notice of the twenty-foot addition that Bradford Chamberlain was putting onto the rear of his saloon to accommodate a billiard table. Railroad station agent Horace Chester (another son of Westford carpenter Alden Chester) and hardware merchant Samuel T. Chase, the paper added on June 22, were putting blinds on their houses, Parsons & Tompkins had recently installed a soda fountain in their establishment, and the Kelley brothers were trenching for the foundation of their new building. Race Street soon was to have walks installed, and Patrick Fagan was—still (or was it finally?)—painting his house. Looking over the village as a whole, the paper counted five new houses going up at present, and it reported that another, this one built for grocer Joseph Swartout, was to be erected soon near a Main Street store building Swartout owned.

At the very end of June, Bradford Chamberlain was said to be ready to install his billiard table in a day or two, while Will Merrihew, the painter, was fencing his yard just north of the Baptist church. As the new month came on, a recent rainy spell had let up sufficiently that several houses were being painted. Mr. Squires's new house—now described as the finest in the village—was presently ready for the plasterers. At the depot, a partition between the two waiting rooms was being removed so as to make one large space, while around the Methodist parsonage near the cemetery a new picket fence was going in. Later in July, there were reports that Bennett's "upper" house was enclosed, that Samuel T. Chase was adding a picket fence to the house he'd recently put blinds on, that the dilapidated school building on the Westford road was to be repaired, that trees had been planted on new streets down near the depot and the church, and that another street had been straightened, displacing several old graves. Timothy Parsons's house was "enclosed, and ready for the joiner work" by July 27, when James Holmes was just framing up the Hubbard house in Westford and preparing to move inside. Milo Kelley, came the news at July's end, had decided to retire from business, having sold his stock to former competitor Jacob Butts (now the village's only cabinet-dealer); the building project of Milo and his brother Lewis accordingly was to be abandoned. But John T. Thompson was still planning to erect his bank, the paper noted, and Lane & Hotchkin was still manufacturing a large number of washing machines, so business continued to seem brisk in Schenevus. Likewise, although Miss E. E. Parrish was closing her millinery shop and moving to Albany, the paper could report early in August that Mrs. S. N. Jones had moved back to Schenevus and would reopen *her* millinery shop in the lower end of the village.[37] Some things balanced out.

Bradford Chamberlain, his billiard table presumably in by August 3, also had been spending considerable amounts of money on his house, which was just a few doors away from his saloon on Main Street. The latest improvement was a "tasty verandah" just then finished. Months after word of his plans first were published, cooper Peter Brady was reported still at work in August making a "respectable house" out of the old shoe shop near the Methodist church. At the same time, Conover and Davis, the new firm now in possession of Schenevus Mills, formerly owned by Egbert Crippen and Nelson Lane, was planning to repair and reopen the complex. By month's end, Joseph Swartout's new house (reported to be in planning in June) was at last underway, the cellar hole having been dug near Dr. Ball's office on Main Street. By then, too, the foundation of John T. Thompson's new bank building was being laid, and plans were being made by blacksmiths Michael O'Brien and William Dutcher to reopen the old Boyce store, where Bennett's tailor shop had been temporarily housed for the summer. By mid-September, when workmen were repairing and painting that structure, construction jobs were trailing off. But Conover and Davis continued to refurbish the mill they had purchased in the summer, reditching the flume so as to increase the water supply and then, by September 21, starting to mill grain. Peddler John Fink, who had been reported in February to be planning a house on a lot south of the depot, at last saw the contractor Austin Tompkins begin the job late in September. And some new work was still in contemplation even this late. Sherriff Harvey W. Brown, the first October issue reported, was contemplating erecting a house just above the village, near his father's, to occupy when his term of office should expire. Wagonmaker Loren T. Brown, perhaps the sherriff's kin, also was planning to build a new residence, this one on his lot on Centre Street. Moreover, hardware-dealer Chase & Mills was erecting a coal storage house near the depot, and a new bridge was being readied for Elk Creek road, where a recent freshet had taken out the old one. By mid-November, Thompson's bank building had been enclosed, and carpenter and builder German Wright, then painting and papering and introducing various internal improvements on his own house, was advised by the paper to "give the outside a brush, . . . and make a clean sweep of it." As it had often during the year, the paper used its pages not only to record but to urge change as well, change that would accord with its view of the village's destiny. No wonder that at year's end, looking back, it concluded that the past twelve months had been good "for both the merchant and the mechanic."

The work recorded in the *Schenevus Valley News* across the period from August 1868 to December 1869 is indeed substantial. It includes a total of twelve new buildings definitely reported as under construction, three houses in the later part of 1868 and eight across 1869, plus the substantial Baptist church building. These figures are consistent with the eight houses (plus the plaster mill and the cheese factory) that the *Freeman's Journal* reported as having been built in Schenevus in 1867 as a whole. A similar comparison unfortunately cannot be made for the calendar years 1868 and 1869 because the Cooperstown paper did not report on the activity in Schenevus in its story of January 1869, and in early 1870 it did not compile any such figures. But it seems safe to assert that across much of the latter part of the 1860s, the village scene was busy with construction work.

It should be recalled, too, that the *Valley News* reported on a large number of repairs, additions, and refurbishings across the seventeen months surveyed. And during this period a good deal else was going on in Schenevus. Division Street, laid out earlier, was extended to the depot, and lots were being sold on it; a new bridge was put in on the Elk Creek road; walks were being installed along Race Street; two other streets underwent substantial repairs; and two roads between the depot and the new church were planted with trees. Although the word may not seem to fit a village the size of Schenevus, the place seems to have been having something of a boom. So caught up in this temporary expansion of business was the village that the paper pointed out in March of 1869 how strong the demand for rentals in Schenevus was at present and urged the local "men of means" to erect cottages or duplexes to rent to an expected influx of mechanics. Otherwise, those workers might not settle in the village because of a housing shortage; at present, the paper noted, "every house and room that can be rented is filled to its utmost capacity."[38]

The industrial background to this local boom was not lost on the *Valley News*. It noted with some excitement that lumbering had been quite active in the winter of 1868–1869 along the Delaware River, a sign of what we may call the industry's regional if not national importance. And in February 1869, the paper was happy to point out that the activity of several people in drawing stone and lumber in preparation for upcoming jobs augured well for the season. The brisk pace at several nearby mills—Winslow and Butler's in Maryland, Barnes and Fox's in Otego and Colliersville, William Ray's in Maryland, and E. E. Ferry's in Schenevus—also drew the paper's enthusiastic notice. Everything seemed so bustling that the *Valley News* waxed very sanguine as spring was breaking over the valley in April 1869: "Almost every week we hear of people contemplating building the next season, and from appearances more will be done in that line this year than last. There is nothing to prevent this place from growing very fast, and also [undergoing] large additions to the business of the town."

But, unfortunately for such boosters, indeed there were many things to prevent such a consummation. A Worcester resident who, in an 1864 letter to the *Monitor*, praised the improvements in Schenevus over the prior decade, went on to enumerate the industrial appurtenances he could foresee coming to that village as the railroad inched closer: an iron furnace and iron works, a trip hammer, an edge tool plant, a woodworking plant, a leather-processing plant, a flax mill, a wool-carding plant and a woolen mill, a brick kiln, and a marble works. By 1869, the village in fact *had* added a brick kiln (although this caught fire in July 1869), a marble works, and a tannery, but otherwise it remained—despite its signs of expansion or the boosterism of its press—a mere country village. As such, it was to find its future role not in the production of goods but rather in the shipping of agricultural produce and the distribution to the rural population of goods manufactured elsewhere. The railroad indeed did link Schenevus with the market economy of the state and nation. But it made of the village in the process a dependent rather than a directive community, one that slowly declined in size and importance over the next decades. When the paper exclaimed in July 1869, "How much better it would be if we had more tanneries, factories, foundries, &c, in our village," it was pinpointing the main reason why Schenevus—which had a better chance to grow than Westford or Decatur or many of the other fifty-odd villages in Otsego County—at last slipped back into a relatively depressed condition. As if recognizing the obvious, the *Valley News* publisher, old optimist Hiram T. Oatman, announced in April 1870 that he himself would be leaving the village for good. Another man was found to take over the paper, young Alden Chester, the Westford native and carpenter's son who had been in Boston lately, sending letters signed "NEDLA" back to the paper from there. But within another few months, he also was gone. Under its third publisher, the paper lasted only one more year.[39]

5 The Carpenters of Westford

Schenevus, whatever its fate several decades later, must have been a good place to practice the carpenter's craft in the late 1860s. It certainly was better than Westford: although James Holmes was lucky enough to have a full year's employment in Westford in 1869, most of the other carpenters there must have been drastically out of luck themselves. Milton Wright, who had been active in the craft in Westford since the early 1840s, left his native village in 1869—almost certainly for the opportunities that Schenevus could offer. With his son William also in Schenevus, and with two of his daughters there as well, married to the two partners in the sash-and-blind factory—a business his own family had been much involved in in Westford—Wright was at the center of a craft circle he no longer could enjoy in Westford. As the discussion of Schenevus in chapter 4 suggests, there simply were many more jobs there. Furthermore, the railroad gave Wright a kind of working range he never would have had in Westford, since he might easily commute to rather distant jobs now on a daily basis. Whether Wright took advantage of this new range isn't clear. All that is discoverable about his work in his new home is that at the end of 1869 he was at work not in Schenevus itself but a few miles away in the village of Maryland, where he labored on the new two-story house that Silas Rider was having built there.[1]

As remarked earlier, the removals of Milton and William Wright and their kinsman Judson Winans typify the abandonment of Westford by carpenters in the period following the Civil War. The craft's representation in the decade following their departure steadily declined, as table 4 makes clear. After recovering somewhat from the effects of the war, which had halved the total number of men in all woodworking crafts, the number fell in the 1870s from its initial high of eleven to six and then to three, until the 1880 federal census reported just one carpenter left at work in the township. With the exception of the sudden high in 1855, this downward trend was a general fact of life in the craft there from 1850 to 1900, roughly the period of Holmes's career. But it would be wrong to draw too many conclusions about the precise nature of the pool of skilled woodworkers in Westford at any one point from these data alone. For one thing, there was considerable in- and out-migration across the period, as well as a large number of occupational shifts. The eight men active in 1865, for instance, are not simply the

Table 4. Numbers of Carpenters in Various Woodworking Crafts, Westford Township, 1850–1900

Craft	1850	1855	1860	1865	1870	1870n	1871	1872o	1872n	1875	1880	1900
C	12	22	10	6	9	4	2	5	2	3	1	3
CM	4	1										
SBM			4	1								
WM			1		1							
SC			1				1	1	1			
ME			1	1								
P		1			1							
Totals	16	24	17	8	11	4	3	6	3	3	1	3

Craft key: C = carpenter; CM = cabinetmaker; SBM = sash-and-blind maker; WM = wagonmaker; SC = shingle-cutter; ME = mechanic; P = painter. See note for table 7.

Sources: U.S. Census population schedules for 1850, 1860, 1870, 1880, and 1900; New York State Census population schedules for 1855, 1865, and 1875; for 1870n, *The New York State Business Directory, 1870* (Albany: Charles Van Benthuysen & Sons, 1870); for 1871, Andrew Boyd, *Boyd's Binghamton City and Susquehanna Railroad Directory* (Binghamton: H. E. Pratt, 1871); for 1872o, Hamilton Child, *Gazetteer and Business Directory of Otsego County, N.Y. for 1872–3* (Syracuse: Journal Office, 1872); and for 1872n, Andrew Boyd, *Boyd's New York State Directory, 1872–1874* (Syracuse: Truair, Smith & Co., 1872).

residuum of those active five years earlier, and the reduction from sixteen to eight is not to be accounted for solely through deaths or outmigration.

A complete survey of censuses and other sources yields data that flesh out these cautions. A total of sixty-one men identifiable as carpenters, cabinetmakers, or woodworkers of some other sort (sash-and-blind makers, most commonly) lived in the township at one time or another between 1850 and 1900. Because of their considerable geographical and occupational mobility, it will be useful to examine the actual figures in greater detail. Table 5 shows when each of these sixty-one individuals was present in Westford and when his work there at one of the woodworking crafts, predominantly carpentry, can be confirmed. It is immediately obvious that the distribution of their craft activity in the township was very uneven.

Taken together, tables 4 and 5 clearly demonstrate that James Holmes's generation, born in the years from 1820 to 1840, came into maturity just as the number of carpenters that the township was supporting began to constrict. The number reported in 1855, the year when Holmes himself first appears as a carpenter, probably was the highest ever present in Westford.[2] Why so many carpenters were there then is not clear. That year and the one previous were not prosperous in general, and especially not for carpenters. Their urban unions, which had just begun to recover from the bad times following the panic of 1837, were being devastated just then by what one labor historian called "the depression of 1854–1855."[3] Perhaps this high number reflects simply the spilling of unemployed urban craftsmen out into the countryside in this "depression," especially since, as we shall see, a fair number of the 1855 carpenters in Westford were neither natives nor longtime residents. But if they indeed came to Westford because they could not find work elsewhere, it is unlikely that they found much employment there then, either. There is no obvious local evidence—such as a major job in progress in 1855 in Westford village, for instance—to suggest that several of the carpenters then present had been drawn there on a specific promise of work. Certainly none of the bigger structures in the village were just being erected then: of the four church buildings standing in the village at the time, one had been built in 1809

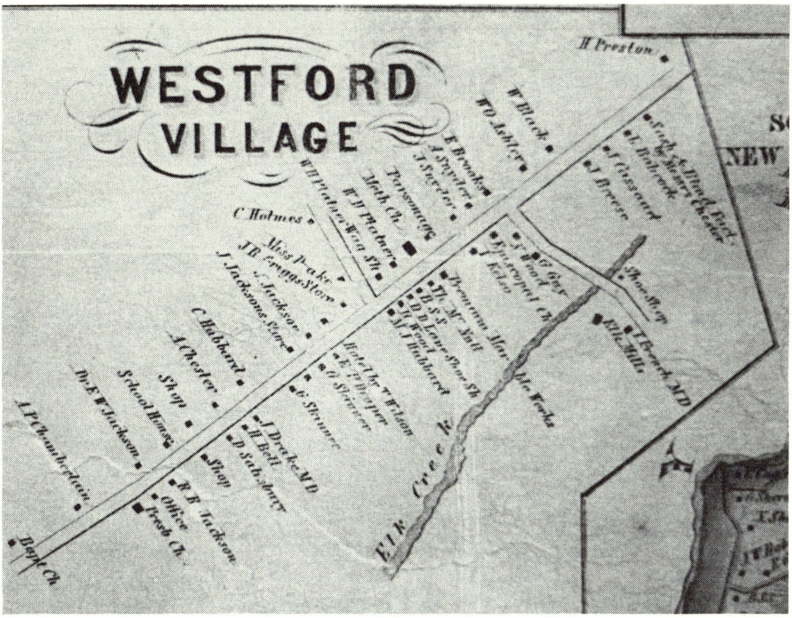

43. Map of Westford village, 1856

and repaired in 1852, one had been built in 1823 and was to be replaced in 1863, one had been built in 1836 and was to be remodeled in the mid-1860s, and the fourth was built in 1840 and would be repaired in 1860. The other church edifices in the township, two in Westville and one a short distance north of Westford village in Bentley Hollow, also had been built prior to 1855.[4]

Since the earliest published map of the village dates from 1856 (fig. 43), it is not possible to determine precisely how many other sizable structures were built there during the early 1850s as a whole, let alone in the census year ending in early June 1855. Still, a brief consideration of the map's information and that contained in the village map in the 1868 county atlas can shed some light on the deployment of the village in this general period, and by inference one may draw some conclusions regarding construction activity in earlier years. The total number of structures shown in the 1856 map, 51, suggests that there had been some growth in the village by the time the 1868 map had been made,

Table 5. Carpenters Active in Westford Township, 1850–1900

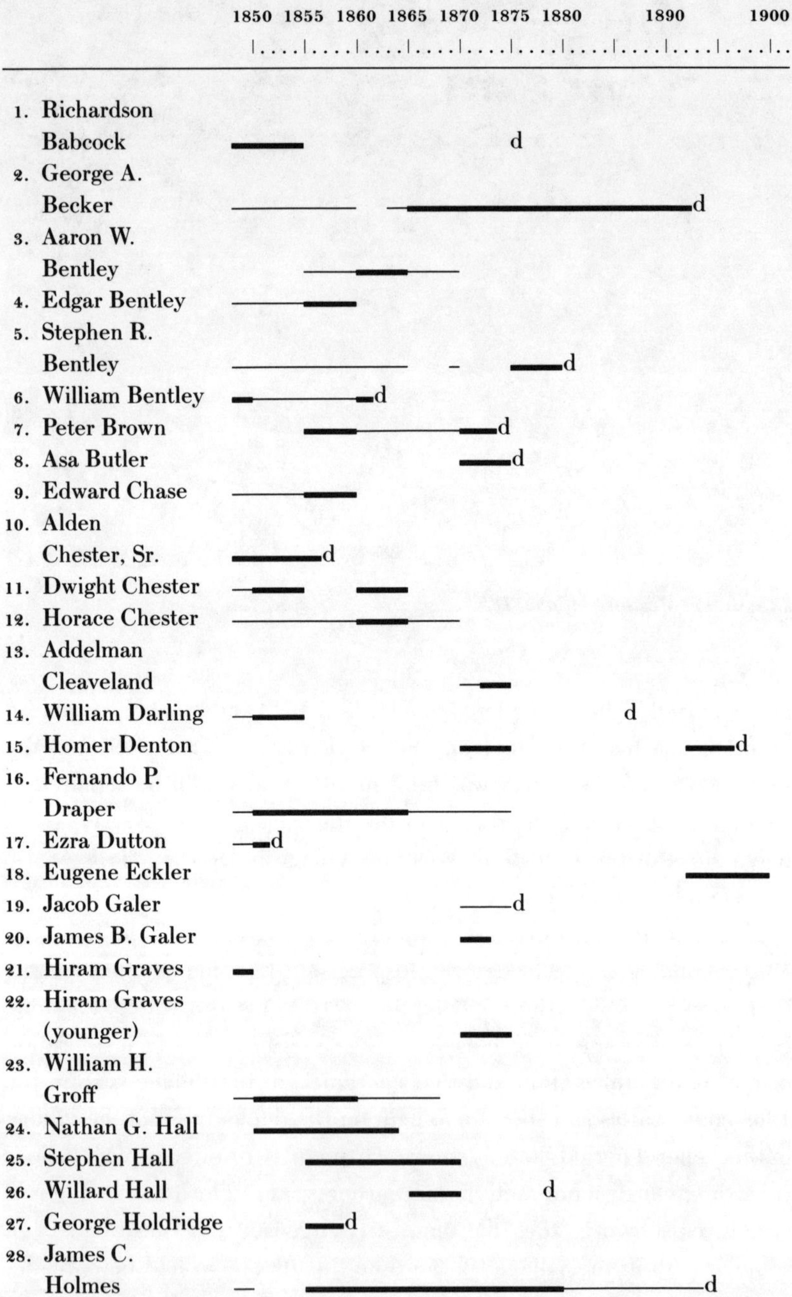

Table 5. Continued

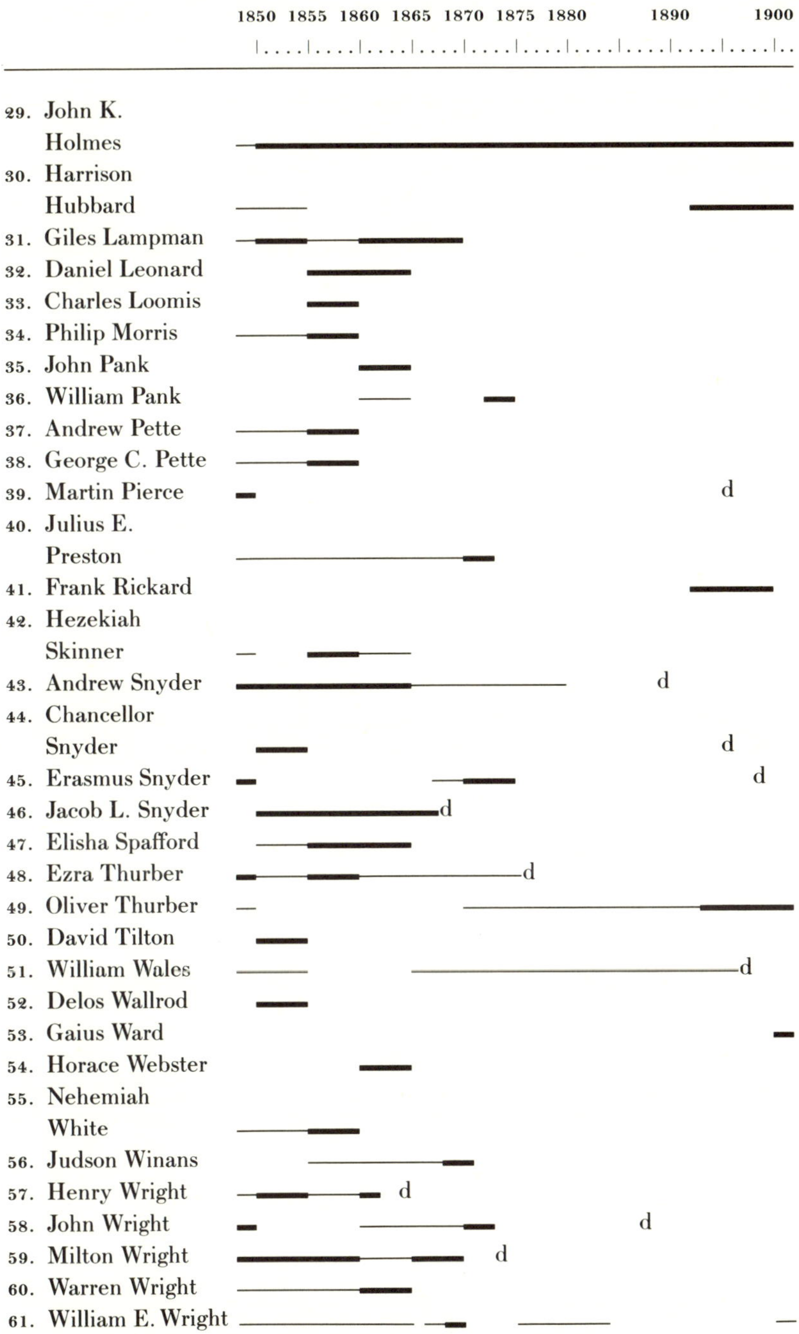

Table 5. Continued

Key: ——— indicates presence in Westford but craft work not confirmed; **▬▬▬** indicates at work at craft in Westford; d indicates year of death when known.

Note: The primary crafts included here are carpentry and cabinet work. Individuals at some time listed at either of these crafts in Westford are considered active at other times if listed then as engaged in such secondary crafts as wagonmaking, shingle-cutting, or sash-and-blind manufacture; some individuals otherwise strongly associated with known Westford carpenters are included here even if they are known to have practiced only the secondary crafts in Westford. Two individuals shown on the table but without active years in Westford (William Wales and Jacob Galer) are known to have practiced carpentry elsewhere at other times and hence are here included; two whom the census did not list as carpenters in Westford (William E. Wright and Judson Winans) but whom the Holmes diary shows to be at work in the craft there are listed as active in the late 1860s. Two individuals (Addelman Cleaveland and William Pank) identified as carpenters in Westford only in the 1872 Otsego County Directory are here listed as active for 1872 and the two years following.

The terminal dates shown here for any carpenter, as in table 4, are not in every case exact. Usually, but not always, a period of activity that ends in the year prior to a census (e.g., 1859, as in the case of William H. Groff) merely indicates that the worker in question was listed in the previous census but not in the next one; that is, he is given the benefit of the doubt either as to his continued presence in the township, or his continued work there at his craft, or both, up to the fifth year since the previous census. Terminal dates that fall in other years are more exact; these derive from sources other than the census, examples of which will be cited in my text at the proper time. But in some instances such other sources indicate that a worker left Westford in a year prior to a census: Erasmus Snyder, for instance, is known to have joined the Westford Baptist church in February 1872 and to have been chosen a trustee of it twelve months later. Twelve months later still, the congregation chose a replacement for him because he had removed to Middlefield, where he had lived earlier and where he was to be buried in 1898. When necessary in the text proper (and in appendix 2), such cases are distinguished from those in which no information other than the census is available.

Sources: Decennial years mark reliance on the U.S. Census population schedules for Westford. Mid-decade years in the timeline indicate use of New York State Census population schedules; there was no 1885 or 1895 state census. There were minimal censuses taken by the state in both 1880 and 1892, which I have employed here to fill out other data, especially since the 1890 U.S. Census is of course not available for use. I have also consulted the directories referred to at the bottom of table 4 to compile this table.

since that plat shows 67 buildings. The latter total includes a number of outbuildings apparently ignored in 1856; when those accessory structures are excluded, the 1868 total (64) still suggests that considerable construction had been undertaken in the village in the interim. A building-by-building comparison of the two maps does not yield exact results, but it tends to confirm this impression. Probably 8 and perhaps 9 new houses were added to Westford in this period, bringing the 1856 total of 32 up to 40 or 41. Two buildings appear to have been added to the mill complex, in addition, while 3 shops, 1 store, and a hotel were also built, plus the new Methodist church. Finally, an 1856 shoe shop was converted into a house, the sash-and-blind factory became a flax mill, the H. Preston house was made over into a cheese factory, and the old Methodist church was moved across the street to become the home of the Westford Literary Institute. Finally, it is worth noting that 3 buildings seem to have been lost in this decade.

The village thus grew by 16 or 17 buildings from 1856 to 1868, an increase of about one-third in just twelve years. But even if proportionate growth had been experienced in the few years prior to 1856, there would have been little reason for the carpenter population in Westford to expand as much as it did from 1850 to 1855. Holmes's experience in 1869 suggests that far fewer carpenters could easily have handled the largest estimate of jobs completed then. There clearly was a considerable surplus in 1855.

Some additional information may help to clarify this picture of the township and its chief village during the period. Other counts of buildings in Westford village exist but are likely to be unreliable. The first, entered in the 1855 manuscript census, compiled in June of that year, states that Westford village contained 40 houses at that time, a number not exactly consistent with the total of 32 shown on the map made the following year. Likewise, an 1860 gazetteer is as far from the mark in the opposite direction when it states that Westford village contained 2 churches and 12 houses at that time. Population figures for the village are also of dubious accuracy at times, since Westford village remains unincorporated to this day and hence never has had any official boundaries. An estimate given in Williams's *Annual Register for 1840* places the figure at 120, whereas the 1855 state population schedule contains in its "Remarks" a probably much more reliable figure (it was entered there by Aaron Pette, a local resident) of 181. The 1865 census raises the number to 210, and the 1872 county directory rounds this off to "about 200."

While these figures are softer than one might like, they may suggest—especially in conjunction with the information to be derived from the two maps discussed above—that at least some of the losses in township population from 1840 to 1865 (a drop from 1,478 to 1,282) were accompanied by a further internal shift of population off the land and into the village, or rather villages, of Westford township. Such a shift would account for a housing market that was expanding even though the town's population was falling, as seems to have been the case in Westford at midcentury. Various census schedules, for instance, show an increase in occupied dwellings even while reporting declines in overall population. The figure for occupied dwellings did dip between 1860 and 1865 (from 292 to 285), but it had risen to the former level from 276 in 1855, and between 1865 and 1870 it again rose, from 285 to 304. Since the total number of houses in 1860 (including 21 unoccupied dwellings) was 313, and the total ten years later (including 15 unoccupied) was 319, the township must have added at least 6 houses in the decade, and of course this figure in practice must have been higher due to loss by fire, collapse, demolition, and conversion of dwellings to outbuildings, shops, and the like. This was, at a bare minimum, a measure of what had been done in the township in those years, a great deal of it centered in fact in Westford village.

Even though there were no major jobs underway in Westford in 1854–1855, jobs so large that they might have drawn in outside workers, the 1855 census does confirm that many carpenters then present in Westford were only marginally associated with the community. One notes first, by way of background, that the holdover from the previous census, when 16 woodworkers were counted in the town, was rather low. That is, of these 16 men, only 7 remained in the township as carpenters a mere five years later. Two others were still present but were no longer listed in the trades, while another 6 had left the area in the interim and one had died. Among the 24 men counted in 1855, there thus were 17 who had not been counted among the woodworkers there in 1850: 16 of these 17 were wholly new to the crafts in Westford, and 5 of these 16 were completely new to Westford as well; the 17th worker previously had worked as a carpenter in Westford but had been out of the trades for an interval. Only slightly more than a quarter of the carpenters active in 1855, in other words, had been in the trades in Westford a mere five years before. Moreover, the average length of residency in Westford for these 24 men was really quite short. Since the state census that year asked how many years each person listed had lived in the township where he or she currently resided, we can easily determine such an average not only for the carpenters but also for the population as a whole. The average residency for the carpenters at that time was 18.5 years, which was about half their average age (36.7); for those entirely new to carpentry in Westford that year, the average residency was 13.5 years, while the average age was 31.8.

This was not, then, predominantly a local craft population emerging from well-entrenched shops but rather a markedly mixed group. Were these men victims of the displacements that modernization was inflicting on the craft, or were they practicing an old itinerancy that could survive past the Civil War in such backwaters as the southern tier? Several of the workers had arrived quite recently, three claiming to have spent only two months in Westford, a fourth a year, and two others two years.[5] By way of comparison with larger groups of the population, one may note the following patterns, which are shown in table 6. In 1855, the group of all men over 21 years old who were engaged in all crafts in Westford yielded an average age of 37.6 years and an average residency in the township of 16.5 years. For the whole of the male population above 21 years—which would include all craftsmen, plus all farmers, all laborers (mostly farm workers), and all tradesmen and professional men, as well as all those for whom no occupation was stated—the average age was 41.5 years and the average residency 21.2 years. For declared farmers alone, by far the largest single group in the township, the average age was 42.8 years, and the average residency was 23.3 years. As a final refinement, it is worth noting that of the 239 farmers reported in 1855, 179 were listed as owning land. The average age of this landowning group was 44.9 years, and their average residency was 25.9 years. Among major occupational groups, the craftsmen

Table 6. Average Age and Residency for Males over 21, by Various Occupational Categories, Westford Township, 1855

Occupation	Average Age	Average Residency	Rank Order of Group by Years of Residency
New carpenters (n = 16)	31.8	13.5	7
All carpenters (n = 24)	36.7	18.5	3
All crafts (n = 48)	37.6	16.5	5
All trades, professions, and other nonfarmers (n = 76)	40.0	17.7	4
All farmers (n = 239)	42.8	23.3	2
All farmers not owning land (n = 60)	36.3	15.6	6
All farmers owning land (n = 179)	44.9	25.9	1
Total of all males 21+ (n = 363)	41.5	21.2	

Source: New York State Census population schedules.

of Westford in 1855, in other words, were the youngest gathering of adult males—except for those farmers who owned no land (mostly renters, we may assume, although some were young men living with their fathers or old men living with their children). And, again except for the landless farmers, the craftsmen had the shortest average residency in the township. Among the craftsmen as a whole, the carpenters were slightly younger than the average but had been in the township about two years longer. Of those carpenters new to the craft in Westford in 1855, the average age was almost six years younger and the average residency just three years less than the averages for all crafts. This last group was slightly more than thirteen years younger than the average landowning farmer and had been in Westford barely half as long.

The state census of 1855 was the only one to ask how many years each person had resided in a given township, so similar calculations cannot be performed in other periods. But it remains possible to describe, by means of other data, some of the other patterns obtaining later. The overlap from 1855 to 1860 was somewhat more substantial than that from 1850 to 1855. But of the 17 men active in the crafts in 1860, only 9 had been active in Westford in 1855. Of these 9, 4 could date their activity uninterruptedly to 1850, while only 5 of the 16 men who had been new to the crafts in Westford in 1855 still were active there in 1860; another 3 who had been active in 1850 but not in 1855 now had returned to work. Hence the persistence over this second period is really not much better than for the earlier one. One notes in both instances that most men new to the trades in a given year were not active in it five years later: 7 out of 12 (58 percent) from 1850 to 1855, 12 out of 16 (75 percent) from 1855 to 1860. The situation remained much the same in the next five-year period. In 1865, there were 8 woodworkers active in Westford. Owing in part to the war, only 1 of the 7 carpenters (14.3 percent) new to Westford in 1860 was part of this later group. The carryover in 1865 from the group that had been new in 1855 was 2, and 3 continued from the comparable 1850 group as well; since 1 of the remaining 3 in fact was returning to activity after a lapse, only 2 men were wholly new in 1865, this too being a likely effect of the war. In 1870, only 3 of the total of 11 craftsmen then at work were carryovers from earlier periods, 1 each from 1850, 1855, and 1865. Two other men had resumed careers interrupted since a decade or two earlier. The rest, a total of 6, were now counted for the first time, even though it is clear from various manuscript records that 1 of them, Erasmus Snyder, also was returning to the township (and to his craftwork there) after a considerable absence elsewhere.

Four of the 6 new craftsmen in 1870 were entirely new to Westford as well, but by 1875 this revived importation of labor had virtually ended. Of the modest total of 3 workers reported then, 2 were carryovers—1 from 1855, 1 from 1865—and the 3d was a lone in-migrant, and actually a returning native son at that. Five years later, this lone new worker had died, the man who had carried over from 1855 (this was Holmes) had now become simply a farmer, and only the 3d of the 1875 trio was left to ply the trade in all of Westford. When the next surviving federal census was taken in 1900, Holmes and the other man both were dead; the 3 carpenters listed then included 2 returning natives and a single man wholly new to the township. In mere numbers, 1900 thus showed an increase over 1880, but if the trade was not dead locally by then, it hardly could have offered any young Westford native the kind of prospects that James Holmes may have perceived in carpentry in the early 1850s. Indeed, by as early as 1870, merely fifteen years after he had launched himself on his career, Holmes himself was already the only member of the extraordinary new group of 1855 still active in Westford, and 1 of only 2 men from that group of 24 still living there, regardless of occupation.

Starting with 1850 (data from earlier years are spotty and often inferential), it is possible to calculate the years of craft activity in Westford and the total years of residency there, both active and inactive, for all the sometime carpenters of Westford. The questions to be answered are these: how long did the average Westford carpenter live in the township? and for how many of those years did he actively ply his craft there? The average for total residence in Westford in the years following 1850 for the 61 carpenters listed in table 5 was 13.65 years (the median is 10 years); the same group's average period of activity in the woodworking crafts there was somewhat more than half that total, or 7.28 years (median: 5 years). It should be noted that both these averages are inevitably inflated because credit for being active or merely present in Westford has been given for a full 5 years following each census listing that is not confirmed by the next one, except when it is positively known from other sources that an individual had stopped practicing or had left at some definite earlier date. (For example, Edgar Bentley is listed as a carpenter there in 1855 but is not present at all in the 1860 listing: still, he is counted as present and active through 1859.)

There is no way of eliminating this inflation. But if everyone credited with a full 5 years' activity following a single census entry were to be credited with only about half that time, 3 years, the average active career of all 61 carpenters would fall to 5.74 years (median: 3 years), a significant decrease. By similar means, the average residency for the 61 men might be reduced to 12.9 years (median: 8 years), with the result that the proportion of activity to residence would drop. Whichever numbers are used, what they suggest is that for 50 years in Westford those practicing carpentry spent somewhere around a decade in the township, and they actively pursued the craft there only about half as long at most. One senses again the extraordinary instability of carpentry in Westford over this period, so at odds with the relative endurance of James Holmes in the vicinity.[6]

A TURN TOWARD BIOGRAPHY

Two other adjustments may help refine these conclusions further. First, one may probably assume that the average Westford carpenter did not intend—and in any case almost certainly could not have managed—to practice his craft full-time in any given period. That is to say, the passing into and out of activity reflected in the five-year intervals of the census may capture a certain enduring indeterminacy in the craft, especially in rural areas. There was a tendency for any carpenter to engage over long stretches of time in the sort of piecemeal economic activity that Holmes's 1869 diary shows him engaging in, on a more local level, even during what was for him a very good year in the craft. What W. J. Rorabaugh writes of the rural "farmer-craftsman" in colonial New England—that he mixed "agriculture and part-time craftsmanship"—seems to have been true late in the nineteenth century in rural New York, much as it was true in the world of Alabama sharecropper Nate Shaw in the twentieth.[7]

The composite pattern seen in the economic life of carpenter James Holmes in the successful year of 1869, in other words, almost certainly applied across wide stretches of his career and the careers of his fellow woodworkers. Especially among the more marginal workers, carpentry was picked up and then dropped, only to be picked up again, as the local demand for labor expanded, contracted, and expanded once more, either during a given season or across wider intervals of time. A look at the lives of two other Westford carpenters may help to sharpen this issue.

HEZEKIAH SKINNER

Hezekiah Skinner was connected by marriage to both James Holmes and Milton Wright, and thus he was part of a craft network that might have been very important to his own advancement. Unlike both of

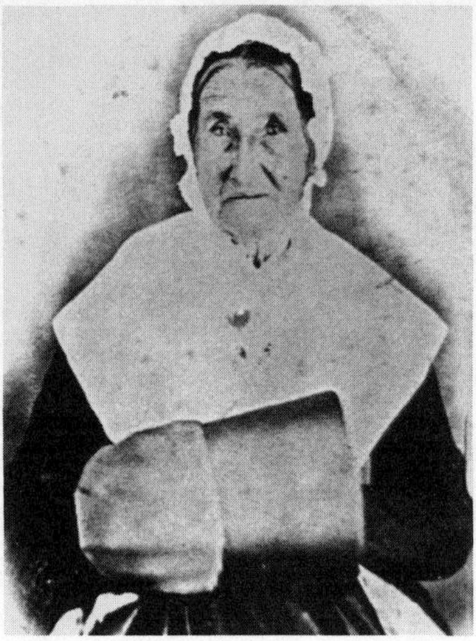

44. "Quaker" John Skinner and Anna Smith Skinner, ca. 1850

those other carpenters, however, Skinner was not a Westford native. His parents, John Skinner and Anna Smith (fig. 44), met just across the Westford border in Seward township, Schoharie County, and were married there in 1808. They had come to this place by devious routes: Skinner, born in Massachusetts, had moved to the Otsego area with his parents around 1801; Smith, a native of Greene County, New York, was herself the daughter of migrating Massachusetts natives who eventually settled down in Schoharie County, achieving there some measure of prosperity. Prior to the wedding, John Skinner had lived in Worcester township, but the newlyweds apparently first resided near Anna Smith's parents some miles northeast of there. In a pattern typical of rural families of the period, they seem to have wandered about the district for a long time, domiciled first in one township and then another, probably renting land on a yearly basis.[8]

It was a scattering life, and their twelve children were born in several different places over the next two decades: Hezekiah, the eleventh, was born somewhere in Schoharie County in 1831. Not until the Skinners purchased property north of Westford village sometime in the 1830s, probably in 1837, did they become really settled. There at last, next to or at least near the house of Calvin Holmes, the Skinners farmed steadily during the following decade, as even the youngest of their children reached maturity. So many of their children eventually married local residents and bought adjoining farms—like James Skinner, who became the husband of Holmes's sister Fanny—that the region where John and Anna had settled down came to be known as Skinnerville.

By 1850, however, several of their other children were gone from Westford, young Hezekiah among them. Like James Holmes, then away studying medicine, Hezekiah presently was dwelling in nearby Delaware County. There he was living with two other males, Clark and Peter Tiffany, aged twenty-one and twenty; Hezekiah, now nineteen, was listed as the head of this obviously temporary household of three young men, all of them practicing carpentry. What brought these workers hither and kept their menage together? The answer almost certainly lies with their next-door neighbor, a thirty-six-year-old housebuilder named Jacob H. Wright. It is unlikely, for one thing, that three young men such as these, especially given their living arrangements, would be self-employed. And it seems fair to assume that Jacob H. Wright's status as a housebuilder meant he had other workers in his employ, workers of precisely their stamp. Even more telling, though, is the fact that Wright was Hezekiah Skinner's kin: he was married to his sister Lucinda. It seems very plausible that young Skinner and probably the Tiffanys as well were not only working for Wright but in fact had moved to their present location in order to work for him. That Skinner just coincidentally showed up in the 1850 census living in this other county, next to his sister and employed in the same trade as her husband, seems all but impossible.

But how, exactly, did Hezekiah come to join his brother in law in Delaware County? It is conceivable that their association dated from an earlier time. Although it may be that Jacob Wright was another of Milton Wright's half-brothers, that seems *not* to have been the case. Still, there are other strong associations between Jacob Wright and the Westford region, not the least being the presence of his wife's family there. In addition, at least two of Jacob's four children may have been Otsego County natives (there are conflicting census declarations on this issue)[9]; and if they *were* Otsego natives, then Jacob and Lucinda Wright themselves would have been Otsego residents from 1841 until at least 1843 and perhaps as late as 1849 or early 1850. By that point Hezekiah was nineteen, and hence he well might have worked for his brother-in-law close to Westford and might have "trailed" him to Delaware County once Wright moved there—or back there, since Wright would seem to have been a Delaware County native. It is unlikely, on the other hand, that Jacob and Lucinda Wright lived in Westford proper during the 1840s. For one thing, Hezekiah's sister, considerably older than he was, seems not to have spent very much time in Westford until after 1855. Born in 1815, she had reached young adulthood before her parents finally bought their Westford farm around 1837, so it is conceivable that she was out in the world on her own several years before then. And indeed she must have married Jacob Wright by 1835 or 1836, for their first child was born in 1836 or 1837.

So Hezekiah, if he had had any contact with Jacob Wright before joining him in Delaware County, probably did not have it in Westford proper. Unfortunately, although Hezekiah was back in Westford in 1855 and hence answered the census question regarding how long he had lived in the township, his answer ("eighteen years") was the same as that given by his parents and others of their children (but not by Lucinda, who said "one year"). Since Hezekiah in fact was *not* in Westford in 1850 but was with the Wrights in Delaware County, his 1855 answer simply must be in error. (There is no reason to suppose that the whole family was gone for a period of time, although perhaps that is a possible explanation here.) If Hezekiah's 1855 figure *is* in error, it cannot be used in trying to determine how long prior to 1850 (if indeed at all) he had been absent from Westford. His absence from the township for several years prior to 1850 would not in itself prove that he was living elsewhere with the Wrights, of course; but it would lend some small circumstantial weight to such an interpretation. As things stand, the truth simply cannot be determined.

How, then, did he come to learn the craft? One possibility is that he was taught it by Milton Wright, who was active in Worcester early in the 1840s and remained active after he moved back to Westford in 1843. This possibility is made more credible by numerous associations among the Skinners and the Westford Wrights, although in fact those mostly came later than the 1840s: Lucinda Skinner's younger sister Hannah would marry one of Milton Wright's half-brothers, Flavel B. Wright, in the middle or late 1850s; and Flavel's brother Warren, in turn, was to marry Lucinda Skinner Wright's daughter Ann in the early 1860s.[10] (Then, too, the Wrights, the Skinners, and the Holmeses all generally lived in the same region of the township, north of Westford village: as is often the case in rural districts, the two main senses of propinquity—of space and blood—shade off into each other.) So it could have been through Milton Wright, then, that Hezekiah, sometime in the later 1840s, first learned carpentry. It seems likely, in any event, that when he first left Westford it was as a young but experienced worker called off to join a kinsman who—as a housebuilder, not merely a carpenter—had need of another hand and had sent to his wife's large family for one. What Hezekiah got in return for relocating was the patronage of an older craftsman who presumably had enough jobs in prospect to employ himself and three hands. In this region, that was nothing to dismiss without serious consideration.

But if this indeed is how things fell out, Jacob Wright's patronage was not to last Hezekiah Skinner for long. For before his fortieth birthday, probably in Delaware County and probably in 1853 or early 1854, Jacob Wright died. By the time the 1855 census was taken a year or two later—the pertinent sheet was filled in on June 12 of that year—Jacob's widow, Lucinda, had established a separate household with her three daughters (the fourth had died in October of the previous year and had been buried in Westford) in the Westford home of her father, John. Hezekiah himself seems to have trailed her back there and was then living in the same house. The Skinner family still was farming; "Quaker" John was now sixty-nine (he would live to be ninety) and probably was being helped in the fields in 1855, as he had been in the past, by his older son, William.[11] But Hezekiah himself, beyond all the changes of the past half-decade, remained a carpenter. Now twenty-four, he in fact was one of the "new" Westford carpenters of 1855.

So how new was he? Perhaps the two years or so that he had spent in his uncle Jacob's employ by 1850 gave him a real claim to that occupational title in the U.S. Census that year. Yet we might just as easily argue that he had assisted Milton Wright in 1845 or 1846 while still in

Westford and so hardly was new there in 1855. The periodic quality of the census thus may cause us to underestimate the continuity of any given worker's career. Hezekiah Skinner can be *proven* to have practiced in Westford beginning only in 1855; but even if he did not practice there earlier, he at least made up part of that larger population that surfaced only partially in any enumeration.

On the other hand, it will not do to overestimate the continuity of any worker's employment in the crafts. Not even when two adjoining censuses give the same occupation for an individual can we be sure that he continuously practiced that trade through the interim. Here again the instance of Hezekiah Skinner may provide an illustration. A plausible case could be made that he remained as much a farmer as a carpenter throughout his early adulthood, true to his immediate family's habits, especially since his next—and last—appearance in the Westford census in 1860 shows him as a landless farmer, married and with one child, living in the general area of Skinnerville. This occupational shift was not permanent; it clearly did not survive Hezekiah's next move, for when he was listed in Worcester in 1865, it was once more as a carpenter. Yet since dual occupational listings in the census are uncommon and inconsistent, we never will know from such sources precisely how Skinner divided his efforts: serially (across the years) or in a more nearly day-by-day manner, with the census listing dependent on seasonal shifts or temporary surges rather than on a programmatic change of occupation extending over a longer time. The dual or even multiple listings one is more likely to find in sources such as county directories suggest just how much of the truth may have been missed by census enumerators. But even where we have much more information, as in the case of James Holmes, a worker's drift over time toward a heavier engagement in agriculture may be hard to factor; Holmes finally in 1875 declared himself both a farmer and a carpenter, and by 1880 merely a farmer, but already in 1869 his diary shows him—still officially just a carpenter, pure and simple—as engaged in a fundamentally mixed economic life.

Surely many of the other carpenters' lives could be similarly adumbrated. With regard to Hezekiah Skinner, who according to the official census record entered the craft in Westford in 1855 and left it by 1860, we might refine the information so as to show him as in all probability involved in the craft, regardless of location, for several more years before 1855, perhaps as many as seven or eight or nine, almost certainly as many as five. We might show him, too, as especially active in Westford from June 1854 until, say, late 1857. Then must have occurred the next big change in his life, marriage. And yet at last we must be aware that Skinner's activity was almost surely not exclusive at any one time, and it was almost certainly not continuous either.

JUDSON WINANS

Judson Winans's career provides a more telling illustration of these problems, since more is known about his activities in between census years. Born around 1839, either in New York or Michigan,[12] Judson was somehow related to the Winanses who in the 1820s were active in the Westford Baptist church: Elder William Winans of the Middlefield congregation, who was among Westford's charter members in 1825; Louisa Winans, who joined the next year and probably was William's wife; William and Louisa's children, Louisa and William, Jr.; and one Russell Winans, who joined by baptism in 1833.[13] Who Judson's parents were is not perfectly clear. But William Winans, Jr., probably was his father: the record does indicate that William's sister, Louisa, was Judson's aunt. In Westford, sometime between 1833 and 1835, Louisa married the oft-mentioned Milton Wright, a son of other founders of the church, who himself had joined it on the same day she had in 1833, also by baptism. It appears that Judson and his siblings—George and Marion (or Mary, although these may have been twin girls)—were brought back to New York from Michigan in late 1854 or early 1855.

Perhaps their uncle Milton Wright resigned the clerkship of the church on November 4, 1854 (effective in January of the following year), because he expected to travel to Michigan to retrieve these orphans and settle family affairs there for his wife. Wright actually may have undertaken a trip west then, although in fact he continued to keep the monthly church meeting records all winter, so any such trip must have been short. It is clear in any case that near the middle of 1855 Wright in fact did travel to the Midwest. On June 2 the clerk's book contains the curt notation, "Clerk gone to Michigan."[14] This could not have been the trip on which he brought the Winans children back, however, since when the census enumerator visited the Wright household on June 9, only a week later, he reported that the children already had spent six months in the town by then. That Wright was recordkeeper for the regular church meeting in early May and for a special one on June 24 suggests that his midyear trip was brief, certainly no longer than seven weeks. Perhaps he went at that time to settle business affairs connected with the Winans estate, things left unsettled when he had gone to Michigan in December, picked up the orphans, and returned with them to Westford, all in the course of five weeks—so quickly that he essentially made the question of his resignation moot.[15]

Whenever and however the children arrived, they all soon were living in the home of Milton and Louisa Wright, just a few doors from Milton's half-brother Chester and his wife, Irena, with whom James Holmes and his recent bride, Mary Eliza Shelland, currently were living. Just sixteen when the census was taken, Judson was awkwardly between families and between social groups, neither a man yet nor—in at least two senses—a child anymore. But soon he was being knit into the community. About a year and a half later, on February 28, 1857, he was accepted as a candidate for baptism in the church, although the ceremony (in "Our Jordan," presumably the still wintry Elk Creek) was postponed until milder weather arrived early in April.[16] By then he must have been working, presumably for or with his uncle Milton.

But Wright himself was occupationally mobile at just this period, so it is difficult to determine what Judson was doing even if we assume his association with Milton Wright. As early as 1858, Wright had acquired land, an eighty-acre parcel worth $1,300, and in 1860 he was listed, for the first time in any census, as a farmer. Yet his later listing as once more a carpenter suggests that he had not in any sense abandoned that occupation. And it is quite likely that Judson already had learned a good deal from Wright, not through any formal apprenticeship but rather simply by being his helper, or by acquiring through mere practice that sort of rural handiness that made any farmer who was serious about his labor somewhat skilled at various crafts.[17]

By 1860, in any event, Judson's condition had clarified in other ways: married and living in his own house with his wife and perhaps his sister,[18] by then he also had acquired an occupation, although an unfortunately ambiguous one. His position was ambiguous because he was listed then as a laborer. The Westford enumerator in 1860 initially distinguished this term in its simple form (presumably meaning someone engaged in the building trades or heavy common labor) from the compound *farm laborer.* He stopped making this distinction after the ninth page, however, although an errant *day laborer* (yet a third term, referring to someone without a steady job) cropped up on page nineteen. Had he maintained such a differentiation consistently throughout the schedule, we might feel confident in placing Judson Winans (who appears later than the ninth page) already within the craft, if only in its lower echelons. As it is, we might see Winans, who did not acquire any land until 1862, as already seriously engaged, beyond his uncle's purview, in construction work; or on the contrary, as not yet skilled in any special sense, as someone who survived by doing day labor in the fields or about the village, the lowest kind of work in the area.

Further data from later years do not completely clarify the situation. As Milton Wright readjusted the balance between his activities in the period from 1860 to 1865, he reduced his land holdings, moved into the village, and once more appeared as a carpenter in 1865. At the same time, Winans was moving in an apparently opposite direction. Thus, Winans held on to the land he bought in 1862, bought another parcel in 1863, and appeared in the 1865 listings as a farmer, living now with his family in the house of his in-laws, Luther and Ann Darling, who also farmed. But the nominal designation *farmer* in the 1865 schedule in fact may have been ambiguous. Nowhere did the enumerator that year use the term *laborer*, for instance: at one point early in his listings he started to, only to cross it out and write in instead the term *farmer.* Since Winans was not given a separate entry in the agricultural census for 1865 (as Luther Darling, though his holdings were modest, was; and as Winans would have been if indeed he had been farming seriously on his own), one can only conclude that he was simply a farm laborer then, probably working for his father-in-law. If this reasoning makes sense, it raises new questions. Was Winans's laboring in 1860, then, also of an agricultural rather than a craft nature? Had he, even as late as 1865, not yet begun to learn carpentry from his uncle?

We can begin to answer such questions by reference to the rather similar experience of two other young Westford workers, Judson's brother George and their cousin William E. Wright. Unlike Judson, who seems not to have been in the army during the war, both George Winans and Will Wright had served, so that when the 1865 census was taken they had only recently returned to Westford. They were living then with Will's parents, Milton and Louisa Wright. Like Judson Winans, they were listed as farmers in 1865, but unlike him they owned no land; none of the three had a separate listing in the agricultural census. Where, then, did the returned veterans work? Milton Wright did still own a modest outlot of twenty acres. But that parcel hardly could have engaged both men full-time, and in any case the agricultural census of 1865 showed it largely was being left fallow.[19] George Winans and Will Wright thus must have been essentially marginal workers who had no land and no specific jobs, farm laborers who were in a kind of occupational holding pattern in Westford. Able to work, they as yet apparently had not found either an occupation or even permanent, promising employment. Soon both men were to leave the township; indeed, by 1870, Judson and his brother, and Will Wright, Will's parents, and two of his sisters, were all gone.

What is clear is that when Will Wright left in 1869, he already had begun to practice his father's craft, and so had Jud Winans. We have as evidence of this change in occupation the Holmes diary itself, which records the work of both young men in that same year. It thus is clear that before the decade was out, regardless of their previous occupational listings, these two farmers in fact had acquired enough skill in carpentry to turn toward that craft to support their families. In Winans's case, we know that he was practicing at least as early as 1868, when his own house had been begun. If, as seems possible, he had built and sold at least one house "on spec" prior to 1868, the date of his beginning would be even earlier. We also know that he called himself a carpenter when the canvaser for a new state business directory visited Westford sometime in 1869 or very early 1870; that he appeared in the census that year, in Oneonta, as a carpenter, too;[20] and that when his name appeared in the county directory of 1872, again for Oneonta (technically, for "West Oneonta"), it bore the same occupational tag. For this four- or five- or six-year period, in other words, we have strong evidence of a serious involvement in the craft, so serious that it seems likely Winans sold his new house in Westford shortly after it had been built so that he might pursue his occupation more rigorously in the suddenly booming village of Oneonta. Regardless of the fact that we know nothing of his work in the period of the late 1850s, when he was most closely associated with Milton Wright; regardless of the fact that his status as a laborer and a farmer in 1860 and 1865 may actually refer to his quite menial labor as, in effect, a farmhand; regardless, finally, of the fact that he never was listed as a Westford carpenter except in a statewide directory issued after he probably had left or at least had decided to leave the town; regardless of all these facts, the Holmes diary and the information gained by following Winans out of Westford to his new home in essence reclaim for the record a carpenter who otherwise would have been merely a mysterious name in the 1870 directory. Here, too, the slim official record may be adumbrated. As a result, our sense of all such data is made more complex, more multidimensional: if we are less likely to view a given career as relatively stable over time, we also are less likely to view simple, single facts as either single or simple. When did Judson Winans "become" a carpenter? When, for that matter, did James Holmes cease to be one?

CARPENTRY IN WESTFORD BEFORE 1850

The collective portrait sketched in table 4 may be refined by extending its scope back before 1850, when the U.S. Census first asked for detailed occupational titles. Twice before, in 1820 and 1840, the census had made queries regarding occupations, though in both these cases the results were stated only as household totals and only according to very general categories. In the earlier year, the three categories were agriculture, commerce, and manufacturing; in 1840, when several other categories were added, these three were slightly modified to agriculture, commerce, and manufacturing and trades. In both years, the enumerator was to record how many people in each household engaged in each activity. Because the categories reflected occupation rather than actual employment, the numbers were likely to reflect longterm status as well as (or in some cases instead of) present activity.[21]

In 1820, Westford listed a total population of 1,526. Its manufacturing population included 49 individuals in 36 households. Twenty years later, the population had fallen (after its 1830 high of 1,645) to 1,478; the manufacturing-and-trades category in 1840 contained 68 individuals in 52 households. Ten years later still, in 1850, when the census enumerator entered the specific "Profession, Occupation, or Trade of each Male Person over 15 years of age," the total Westford population was 1,423; the total number of males who probably can be considered as falling into what in 1840 had been the manufacturing-and-trades category included 55 individuals living in 40 households.

In 1850, for the first time, we of course can note the overall breakdown into various crafts or trades. It is possible to use this information as a basis for roughly estimating how many men were likely to be involved in specific crafts at earlier times. Table 7 gives these figures by category for 1850, 1855, and 1860 so that such estimates can be made with a bit more accuracy for the pre-1850 period. The particular distribution among the more important crafts—including carpentry, shoemaking, and blacksmithing—changed rather dramatically over this decade. But the proportion of all craftsmen to the whole population remained relatively stable: 1:28.5 in 1850, 1:24.5 in 1855, and 1:26.1 in 1860.[22] For the pre-1850 period the ratios were about the same, 1:31.3 in 1820 and 1:21.7 twenty years later. Interestingly, carpentry was the most numerous craft in 1855 and was tied with shoemaking for that distinction in 1850 and 1860.[23] Carpentry also was one of only three crafts or trades (the others were mason work and milling) that experienced an increase from 1850 to 1855. This was a period when the crafts were notably thinning in Westford—a point the comparative statistics make clear—but carpentry (and even more so, the woodworking crafts as a whole) remained stable or grew.

Is it possible to estimate how many of the men listed as in manufacturing and trades in 1820 or 1840 were carpenters? Two approaches to

Table 7. Comparison of Distribution among Crafts, Westford, 1850–1860

Craft	1850 (P = 1,423)	1855 (P = 1,371)	1860 (P = 1,382)
Carpenter	12	22	10
Shoemaker	12	9	11
Blacksmith	11	6	4
Wagonmaker	5	2	4
Cabinetmaker	4	1	
Sash-and-blind maker			4
Cooper	3		3
Harnessmaker	3	2	1
Mechanic			3
Miller	2	3	2
Painter		2	
Flax-dresser		2	
Millwright			2
Mason	1	3	3
Ropemaker	1		
Tailor	1	1	
Stone-cutter		1	1
Gunsmith		1	1
Butcher		1	1
Shingle-cutter			1
Basketmaker			1
Green-tow manufacturer			1
Totals	55	56	53

Note: Figures include all workers listed in each craft; totals may differ in some cases from those given in table 4 because the latter counts under subsidiary crafts only those men identified as carpenters at some point in their careers.

Sources: U.S. Census and New York State Census population schedules.

this problem may be suggested, one statistical and one biographical. The proportions of total craftsmen to nominal carpenters in 1850, 1855, and 1860[24]—4.6:1, 2.5:1, and 5.3:1, respectively—vary too much (and in any case the raw numbers are too small) to provide a reliable guide to the earlier periods. Arranged in order from the fewest to the most carpenters (1860, 1850, 1855), they would suggest that in 1820 there were between nine and nineteen carpenters in Westford and that in 1840 there were between twelve and twenty-seven. Even if one accepts the lower of these figures as a guide, success in identifying actual carpenters from the documentary sources available is rather modest. Although it seems safe to assume that in either decade there were at least a dozen men working as carpenters in Westford, exactly who these men were is very hard to say.

As a consequence, a biographical approach to the issue must be very restrained. Of the sixty-one men who are known to have practiced in Westford after 1850, only eight are known to have practiced there before that year as well: Richardson Babcock, William Bentley, Alden Chester, Sr., Andrew Snyder, Erasmus Snyder, Ezra Thurber, John Wright, and Milton Wright; two other men not present at all in Westford in 1850 or later, but listed there in the manufacturing-and-trades category in 1840 and later listed as carpenters elsewhere—these are Hiram Graves and Martin Pierce—ought to be added to this small group.[25] Among the remaining fifty-one carpenters, thirteen mostly younger individuals (table 5, nos. 14, 16, 17, 23, 29, 31, 34, 42, 46, 49, 50, 52, and 57) may have worked in the craft in Westford during the 1840s, although of this group only five (nos. 14, 16, 23, 49, and 57) seem safe bets. Two other individuals (nos. 33 and 44) might be called marginal Westford carpenters during the same decade. That is, both were living in Westville, the village shared between Westford and Middlefield townships, and were active there during the 1840s before they moved over into Westford proper, where they were counted as carpenters in 1850 (no. 44) and 1855 (no. 33).

By way of summary, we thus might figure that of the sixty-one Westford carpenters, somewhere between ten and twenty-five can be counted as having practiced there before 1850. If all twenty-five indeed were active, most would have been active for only two or three years preceding 1850. One, Erasmus Snyder, probably had become active in the earlier years of the 1840s. Of the rest, Babcock, Chester, Graves, Pierce, and the two Wrights could trace their work back to the 1830s. Only three—Bentley, Andrew Snyder, and Thurber—definitely were active before 1830: these three, in fact, were active before 1820 as well. As a final point, it should be noted that a total of eleven individuals active in manufacturing and trades in Westford in 1840 lived in the households of the seven carpenters known to be practicing then. The other four men thus counted probably are additional carpenters not otherwise known. They all were resident with John Wright, who may have been a wagonmaker then in Westford, as he was in Worcester in 1850. But John Wright was never consistent in his occupation, so that issue is difficult to settle. The four unknown workers, in any case, included one under the age of fifteen and three in their twenties.

Beyond this sketchy history of the craft in early Westford there lie only the vague occupational categories of the pre-1850 censuses. The problem is complicated by the mobility of the population, which makes it very difficult to track down workers from the 1820 or 1840 census schedules until the fuller coverage of 1850 reveals precisely what they were doing then and—perhaps—earlier. The great majority of men listed as active in all the crafts in Westford even as late as 1840 simply were not there a mere ten years later. Fully forty-one of the sixty-eight were gone, so the new specificity of the census in 1850 is of little use as a means of reading possible occupations back into the earlier record. Furthermore, a check for both the 1820 and 1840 Westford craftsmen in the indexes of the 1850 census elsewhere (in the rest of Otsego County, other New York counties, and several other states) has turned up surprisingly few among them who later were practicing any craft, let alone carpentry—if indeed they could be located at all.

A glance at the lives of some pre-1850 carpenters will suggest why tracking people down proves so difficult. Richardson Babcock is a good case in point. Born in or near Westford in June 1797, Babcock had moved to Worcester by 1820. The census there showed him as married, and with one child; he and two other young men, one about his own age and the other between ten and sixteen, were engaged in some craft in Worcester at the time. In 1830, he was living in Albany, some seventy miles away, with six children and probably the same wife (named Mary, she had been born in Ireland in 1794), but his occupation of course was not then reported. By 1840, he had moved back to Westford with his family (one daughter was gone now), and was again listed as engaged in some craft. Ten years later, for the only time in his many years actually passed in Westford, Babcock showed up as a carpenter, only to leave again before 1855 and to surface in Maryland township in 1860, again as a carpenter. In 1870, he once more was reported in Maryland but as a farmer, although in the 1872 county directory he made a last appearance as a carpenter. He died in 1875 and was buried alongside his wife in the Schenevus cemetery. Except for his one listing as a carpenter in Westford, he almost certainly would have dropped from the record entirely.

Almost certainly, too, a number of other carpenters practicing in Westford in the pre-1850 period left the region as Babcock did but did not return later and thus *have* fallen from the record. Such might have been the fate of Hiram Graves the elder, who (as explained in note 25) has been included in the master list of sixty-one Westford carpenters even though he never was directly identified as a carpenter in the township. Actually, his only listing in the Westford population schedules came in 1840, when his inclusion among those engaged in manufacturing and trades in itself tells us little. But later listings elsewhere make it plausible that Graves was a carpenter in Westford during the 1830s and 1840s. In 1850, he thus was listed as a carpenter in Oneonta, and by 1860 he was practicing that craft in his new home, Laurens township. It seems fair to assume that while in Westford in the 1830s and 1840s Graves had plied the same craft, although of course it is possible that such an assumption, even if fair, is wrong. At least Graves was the only adult in his household in the 1840 Westford census to whom the manufacturing-and-trades label, however vague it is, conceivably could apply.

The case of Andrew Snyder (fig. 45), an important Westford carpenter from before 1820 until the 1850s and 1860s, entails some initial caution on just these grounds. Snyder had come to the area as early as 1805, and he may have been the one member of his mother's household who as early as 1820 was engaged in some craft. Because there were two young men between sixteen and eighteen in that house then, however, and only one was engaged in manufacturing, there is no way of knowing that it was Andrew who was indicated—let alone that the craft in question was carpentry. Likewise, a similar designation in his own household in 1840 may have applied either to Snyder or to the other male, somewhat his senior, then living there. As a result of such uncer-

45. Westford carpenter Andrew Snyder, ca. 1855

tainties, Snyder might not be credited with practicing in Westford until 1850, even though it seems very likely that he practiced perhaps as much as thirty or thirty-five years earlier. Only because Snyder shows up in Dr. Elhanan Jackson's "Medical Daybook" of the 1830s as a laborer on various jobs—he hewed timber, worked on a school house, "covered" a shed, and made a sleigh for Jackson as partial payment of his bill—do we know with tolerable certainty that he did engage in woodworking that early in his life. As a result of that chance record, he may be safely included among the few early Westford carpenters whose names can be recovered.[26]

The point is that various early cases have been treated in this study as the recoverable context suggests they should be: there are no rules that settle all conflicts. Like Andrew Snyder, for instance, William Bentley spent many years prior to 1850 in Westford and shows up as a listed carpenter at least once after that year. Like Snyder, too, Bentley lived in a household whose adult male population contained some craftsmen, though exactly which of the individuals were engaged in craftwork isn't clear. Bentley was one of three young males living in his own household in 1820, two of them involved in some craft. And by 1840, when Bentley was in his fifties, two young males (one ten to fifteen and the other fifteen to twenty) were then living there; of these three, only two were once again listed as engaged in some craft. Despite this apparent ambiguity, however, Bentley is credited with practicing in Westford prior to 1850 (his only census listing as a carpenter in fact comes as late as in 1860, after two listings as a farmer) because a carpenter's account book documenting his work for the period from 1814 to 1845 survives today in the collections of the New York State Historical Association.

Something might be said here about where those carpenters who weren't Westford natives came from and where those who eventually left the town, whether natives or not, went. As I remarked earlier (note 6), not all those who spent less than their whole lives in Westford have been located prior to or after their sojourn there. But enough have been located for at least part of their years elsewhere to allow for some more general patterns to be sketched. Table 8 shows the probable or known

Table 8. Known or Likely Birthplaces of Westford Carpenters

Birthplace	Number
Westford	23
Worcester	3
Cherry Valley	3
Oneonta	1
Middlefield	3
Other Otsego County	6
All Otsego County	39
Schoharie County	3
Rensselaer County	3
Columbia County	2
Other New York State	3
All New York State	50
Massachusetts	2
Connecticut	1
Vermont	1
New Jersey	1
Maryland[a]	1
New York State or Michigan[b]	1
All United States	57
England	1
Scotland	1
Unknown	2
Total carpenters	61

[a] This could be either the state of Maryland or the nearby Otsego County township.

[b] Judson Winans's birthplace is reported to be either in different schedules.

Sources: U.S. Census and New York State Census population schedules.

birthplaces of the 61 carpenters active in Westford between 1850 and 1900. About 38 percent of them, or a total of 23, were Westford natives, another 26 percent (16) were born elsewhere in Otsego County, and 18 percent (11) elsewhere in New York State. Of the remaining 18 percent (11), 4 were from New England, 1 each from New Jersey and Maryland, and 1 each from England and Scotland. The birthplaces of the final 3 remain in doubt: two were listed as of unknown birthplace in the census, and the third—this was Judson Winans—was variously reported as born in Michigan or New York.[27]

Of more importance than this spread of birthplaces is the degree of persistence of all 61 men, natives or not, in Westford. Of the Westford natives, surprisingly only 1 (James Holmes's brother John) seems to have spent his whole life in the town. To be sure, James Holmes himself and the Westville carpenter George Becker were absent for only one census. Another 5 men who were natives and moved away came back for a period of time later, and yet another 3 came back for good after a sojourn elsewhere. But that still means that only 11 of the 23 natives had at least a moderate degree of association with the township. Of those 11, only 6 actually died there. One senses again the essential pattern of out-migration that ruled the population of this shrinking area. In fact, fully 60 percent of Westford's carpenters were not natives, so the stability of this craft population as a whole was quite low.

In general the other carpenters came from places one would expect. Although the influence of the adjoining townships is rather moderate, that of the county as a whole is relatively strong. Some 42 percent of those not born in Westford thus were born elsewhere in Otsego County. When these and the Westford natives are lumped together, Otsego's share is nearly two-thirds of the 61. The influence of several upstream counties in New York and of the rest of the state is minimal. That of other states is small and focused largely in New England, from which a great many of the first settlers of Otsego County had derived, including the families of both James Holmes and his wife, Mary Kelso. The foreign contingent is very slim indeed, especially considering that the period under study extended up to 1900.

Wherever they came from, these non-natives hardly exhibited a high degree of local attachment to Westford. When the known relocations of all the carpenters over their whole careers are listed (table 9), one can see that many of these men in fact shifted around frequently. The average number of known residences for the whole group (just over three) is small because for the majority of these men it has been possible to reconstruct far less than the whole career. It simply isn't known how many other times before or after or even in the midst of their period of traceable activity they relocated, or how widely they moved about.

Table 9 displays the information in a schematic manner: in a series of groupings by origin, with subgroupings by number of moves and by all later relocations and ultimate destination (if known), it shows precisely how these men tended to deploy themselves in the landscape of Westford and its various environs. As noted above, the attachment of individuals to Westford was not notably strong for the group of sixty-one men as a whole. When the data are broken down as in table 9, it becomes clear, as we would expect, that the natives' attachment was measurably stronger. In addition to the one man who was a life-resident (no. 29), the one (no. 28) who was virtually a life-resident, and the four others (nos. 2, 5, 30, and 49) who returned to Westford permanently after sojourns elsewhere, there were five men (nos. 1, 16, 45, 58, and 59) strongly enough attached to their birthplace to return there after time elsewhere. These five last men ultimately left Westford once again and did not return again. But their association with the township was such that for long periods of time before their ultimate departures they basically shifted back and forth between Westford and immediately neighboring townships.[28] Although they did not remain permanently in their native township, then, it did become a kind of geographical center around which their lives circulated. And when they finally left Westford, without exception they simply moved to another adjoining township: three to Maryland, one to Worcester, and one to Middlefield. For all practical purposes, then, these natives might well be counted as loyal sons. A sixth man (no. 12) who left Westford for Maryland and then spent time in Worcester before returning to Maryland also might be included within the same group, and if the one doubtful native (no. 39) in fact was born in Westford, then his last residence—Roseboom—would throw him into the group, too.

The habit of moving to adjacent townships also was shared by men

Table 9. Birthplaces and Relocations of Westford (WFD) Carpenters

Carpenter Number[a]	Known Domiciles
1. Westford Natives	
29	WFD
28	WFD—Delaware Co./Delhi twp.—WFD
49	WFD—Maryland—WFD
2, 5, 30	WFD—?—WFD
1	WFD—Worcester—Albany—WFD—Maryland
16	WFD—Maryland—WFD—Worcester
45	WFD—Middlefield—Cherry Valley—WFD—Middlefield
58, 59	WFD—Worcester—WFD—Maryland
12	WFD—Maryland—Worcester—Maryland
11	WFD—New York City—Boston
57	WFD—Virginia [with army]
37	WFD—Roseboom—?
40	WFD—Maryland—?
47	WFD—?—Maryland—Worcester—?
4, 9, 38, 55, 60	WFD—?
39[b]	?—WFD—Roseboom
2. Natives of Other Specified Otsego County Townships	
19	Middlefield—Cherry Valley—WFD
61	Worcester—WFD—Maryland—WFD—Binghamton—Worcester—WFD—?—WFD
44	Middlefield—WFD—Middlefield
20	Cherry Valley—WFD—Oneonta
35, 36	Cherry Valley—WFD—?
15[b]	Middlefield(?)—WFD—Maryland—Schoharie Co./Seward twp.—WFD
8, 13[b]	Worcester(?)—WFD—Worcester

Table 9. Continued

Carpenter Number[a]	Known Domiciles
3. Natives of Otsego County	
27	?—Maryland—WFD
3	?—WFD—Oneonta
24	?—WFD—Laurens
26	?—Hartwick—Middlefield—WFD—Laurens
18	?—Roseboom—?—WFD—?
25	?—WFD—?
53	?—Oneonta—WFD—?
4. Natives of Other Specified Counties in New York	
46	Rensselaer Co.—WFD—Middlefield—WFD
6	Columbia Co. (Canaan twp.)—WFD
23	Rensselaer Co.—WFD—Minneapolis, MN
14	Columbia Co.—WFD—Decatur
31	?—Schoharie Co.—WFD—Unadilla
32	?—Schoharie Co.—Cherry Valley—WFD—Middlefield—Otsego
42	Schoharie Co.—WFD—Delaware Co. (Tompkins twp.)—WFD—Worcester
43	Rensselaer Co.—WFD—?
5. Natives of New York State	
17	?—WFD
22, 54	?—WFD—?
6. Natives of Other States	
10	New London Co. (Groton twp.), CT—WFD
48	Vermont—WFD
21	Massachusetts—WFD—Oneonta—Laurens
33	Egremont, MA—Middlefield—Maryland(?)—WFD—Lenox, MA
34	New Jersey—WFD—?
50	Maryland (State)—WFD—?

Table 9. Continued

Carpenter Number[a]	Known Domiciles
7. Natives of Other Countries	
7	Scotland—Philadelphia—Cherry Valley—WFD
51	England—WFD—Cherry Valley—Middlefield—WFD
8. Nativity Unknown	
56[c]	?—Michigan—WFD—Oneonta—Worcester
41, 52	?—WFD—?

[a] Table 5 gives the names of the carpenters listed below by numbers.
[b] Birthplace uncertain but likely to be as indicated.
[c] Birthplace listed as both Livingston County, New York, and Michigan in separate census returns.

Sources: U.S. Census and New York State Census population schedules.

who finally made more decisive breaks with Westford. Three (nos. 37, 40, and 47) who ultimately disappeared from the local record thus made their first moves to Maryland or Roseboom. Only two (nos. 11 and 57) who were traceable appear to have left Westford for rather distant points. Dwight Chester moved to New York City and then to Boston, though like his half-brothers Alden and Horace (no. 12), he probably first moved to Schenevus or Worcester. Henry Wright left for Virginia as a soldier in the Union forces, dying there at the battle of Cold Harbor.[29] As a final note on Westford natives, it should be observed that five of them (nos. 4, 9, 38, 55, and 60) left the township for undiscoverable destinations and seem never to have returned. Because they were not found in the records of neighboring townships, it may be assumed that they also made decisive breaks with the area. These men rather neatly balance the five "circulating" workers who stayed in the general vicinity.

Among natives of the neighboring townships or of Otsego County in general who at one time dwelt in Westford, a rather different pattern emerges. Only three (nos. 19, 61, and 15) of the nine men in the former group became permanent Westford residents, two of them only after moving away (in one case [no. 61] three times) from there to a variety of other places, including the rather distant city of Binghamton, for intervals before returning. The six others merely spent time in Westford while in transit elsewhere. Two (nos. 35 and 36) went to unknown locations; two (nos. 8 and 13) went back to Worcester, their likely native township, thus becoming part of another circulating job market; one (no. 44), in a similar fashion, went back to his native town of Middlefield; and one (no. 20) passed on to Oneonta. The group of seven Otsego County natives whose home townships are not known, listed in part three of table 9, was dominated by men who seem to have drifted into Westford and then left it for good. Although indeed one of them (no. 27) did remain there once he had arrived, six others merely passed through on their way to other domiciles. Whether these last ever returned to their native townships, of course, is a moot issue here. That at least two (nos. 26 and 27) of them came to Westford via neighboring townships may suggest that they were "circulating," although this is a difficult question to address with so few cases in hand. It is true that, although three of the six who did not remain in Westford left for unknown destinations, the three others (nos. 26, 3, and 24) went to other Otsego townships: the degree of countywide persistence, in other words, was rather high. Was the local labor market countywide in scope? Although the limits of the present study will not allow a reliable answer to this question, it would appear that such was the case.

The other in-migrants fall into five categories: (1) those from known New York counties other than Otsego; (2) those from unknown New York State locations; (3) those from other states; (4) those from other countries; or (5) those of uncertain origin. Of all these men (a total of twenty-two), seven remained permanently in Westford. The group from other known counties in New York provided two of the seven, both of them (nos. 6 and 46) moving directly to Westford, though the second one later spent some time in a neighboring township. Like him, four other men from New York counties other than Otsego also seem to have entered the wider local labor market. But unlike him, none of them ever developed close ties with Westford itself. One of them, indeed, moved only a short distance away to Decatur, staying there for the rest of his life. The other three, however, passed considerably farther afield. One (no. 32) can be shown to have arrived in Westford from another Otsego County township, only to leave for a third and then pass on to a fourth township at later times. The second (no. 31) arrived directly in Westford, then left for another Otsego township. The third (no. 42), who had family ties in Westford, lived there as a youth before leaving for a nearby Delaware County township. He later returned to Westford for a period, then moved to Worcester. In only two cases (nos. 14 and 32) were the moves confined to adjacent townships. The question of a countywide or regional labor market again arises.

Among the other migrant groups, some similar patterning can be found. Only one (no. 17) of the three New York State natives from unknown locales remained in Westford permanently. And in fact he lived there only about four years before dying in 1852 at the age of twenty-seven. Whether he circulated locally prior to arriving or came directly to Westford from more distant precincts of the state has not been determined. No trace of his presence locally prior to his first appearance in the Westford record has been found, and the fact that he joined the Westville Baptist church on the same day as several of his siblings in 1848 suggests that he moved in with his family in the 1840s, although it remains possible that they all had longer local ties. The other two New York State natives (nos. 22 and 54) could not be traced either prior to their arrival or after their departure, so they also seem not to have come to Westford through participation in the local labor market.

The two natives of other states who became permanent Westford residents were, as already mentioned, part of those large New England migrations that saturated Otsego County from the 1780s to the 1830s. One of them (no. 48), born in Vermont in 1792, moved to Westford on his own in the 1820s. The other (no. 10), born just a decade later in Connecticut, came to Westford as a boy when his family migrated in 1813. Although one (no. 21) of the Massachusetts natives circulated about the county and spent only a brief period in Westford, he also seems to have fit this pattern of early migration. He must have left Massachusetts when quite young, since he told the enumerator in 1850 that his place of birth was unknown, and only in 1860 was the state named. The other Massachusetts native (no. 33) conceivably came to New York as a young man. He spent very little time in the vicinity before dropping from the record in 1857. The lone New Jersey migrant (no. 34) came to Westford at the age of forty-eight in 1837; he stayed there until at least 1855 but after that seems to have left. Finally, the young man (no. 50) who listed his birthplace as Maryland (state) in the 1850 census merely flashed across the record in Westford, disappearing as suddenly as he appeared. There seems little chance that he came from the nearby township of Maryland, although a few people with his family name do show up in local records there.

The two foreign-born carpenters, one from England and the other from Scotland, showed a good deal of persistence in Westford. The Englishman (no. 51) arrived in the United States in 1848, when he was twenty-six, and appears to have settled in Westford from the outset. His move across the township boundary to Middlefield in a few years was merely temporary, since he returned to Westford in 1865 and was to die there in 1897. By contrast, the Scotsman (no. 7) arrived in Westford by way of nearby Cherry Valley, where he resided virtually from the time of his naturalization at the age of forty in 1828 until the early 1850s. After moving to Westford in the latter decade, however, he too continued there until his death in 1874. The foreign-born carpenters of Westford thus established the highest persistence rate of any non-Westford native group, although of course they composed too small a group for any real conclusions to be drawn from their persistence. At the very least, they certainly were more stable than several American-born carpenters of vague origins. One among the latter (no. 52) told the enumerator in 1850 that he didn't know where he had been born: only twenty-six, he hardly could have forgotten, but in any case he was gone by 1855. The other man of unknown origin (no. 41) passed so briefly across the Westford scene that he left only a single census entry (and this in 1892, when the state tally asked no questions as to origins) and could not be traced either before or afterward. But even Jud Winans (no. 56), despite his kinship ties in Westford, stayed there for only about fifteen years before seeking his fortune elsewhere. Not surprisingly, the men who were of indeterminate origin thus had the worst persistence rate in Westford.[30]

THE CASE OF CHARLES LOOMIS

A biographical approach to the lives of these migrating craftsmen may help to flesh out the increasingly bare statistics here, as the discussion of Hezekiah Skinner and Judson Winans did above. In fact, that earlier discussion, focused on the issue of occupational mobility, provided some useful evidence of geographical mobility as well. But other examples may be more pertinent here. The first involves a man who barely flits across the records in Westford, Charles Loomis (no. 33). The son of Calvin and Nancy Parsons Loomis, he had been born in Egremont, Massachusetts, on December 21, 1819, and had moved to New York State sometime prior to 1850. In that year, when he showed up in the Middlefield census, he was listed as a carpenter and joiner, aged thirty, who owned $600 worth of real estate. He must have been in the Middlefield area for a minimum of about four years by then, since he had married the Westford native Emily H. Badeau on February 2, 1847, and their first child, Ann, had been born on January 1, 1849.[31]

Emily Badeau's family lived mainly in the south central and southwestern parts of Westford, closer to Westville than to Westford village. But the family had ties in each village. Some members, for instance, had become members of the Westville Baptist church in the late 1830s, but in 1838 Emily joined the Westford congregation by baptism and two of her sisters transferred their membership there (from Westville) as well. Whether this new affiliation reflected a change in residence for one or more of the sisters—were they "working out" then?—is not known. Certainly following her marriage to Charles Loomis about a decade later, and perhaps some years earlier on her own, Emily was to move to adjoining Middlefield township.[32] But that shift hardly removed her from her family's domain. On May 25, 1851, a year or so after the Loomises first were reported living in Middlefield, Emily transferred her membership to the church in Westville. And in any case, Middlefield and Westford bordered each other—Westville, as noted earlier, in fact lay partly in each township—so that "home" remained nearby despite her move. Without kin of his own nearby, Charles Loomis seems to have been knit tentatively into his wife's family, joining the Westville congregation by baptism on February 13,

1853. Now more firmly ensconced in the Westville community, the Loomises appear to have been leading relatively stable lives.[33] They owned real estate, were both members of the church, and had two children now, and Charles presumably was still plying his trade as carpenter and joiner.

But if there actually was any stability, it soon vanished. On June 6, 1854, a year and a half after Charles Loomis had joined the church, their firstborn child, Ann, died and was buried in the Badeau Hill cemetery in southern Westford township.[34] When the next census was taken in 1855, Charles and Emily were living in Westford with their two surviving daughters, Ida, who had been born on September 4, 1850, and Emily or Emma, listed as a year old on the 1855 record. Charles still was a carpenter that year, but in the recent past the Loomises must have fallen on hard times. They had been living in the house of Emily's parents, Amzi and Ann Badeau, since April or May, and Charles no longer claimed to own any land.

Nor was this the end of their troubles. Within five months, on November 31, 1855, Emily herself died and was interred near their first child. Then, a year and a half later, on June 27, 1857, the Westville congregation excommunicated widower Charles Loomis. Although the reason is not stated, the action was not uncommon among the Westville Baptists, whose records show that they took the same action against more than sixty communicants—in a smallish congregation—between 1830 and 1900. Charles Loomis, who shows up in no surviving Westford census up to the century's end, apparently returned to Massachusetts rather soon after his expulsion from the church. By 1860 he had remarried in Lanesboro, and in 1908 he was still residing in Lenox. Emma, the younger of his daughters, stayed behind in New York for a time. She was living with her maternal grandparents in Westford in 1860, with her maternal uncle James there in 1865, and as a domestic servant with the family of William Skinner in 1870. By 1880, however, she too had moved to her father's native state, where in that year she was married. Oddly, when Charles died in 1912 he was interred not in Lenox but rather next to his first wife (by then he had had three) and their first child in Badeau Hill cemetery, Westford. This was the most farfetched of any Westford carpenter's "return" to the township.

It is not clear that the briefness of Charles D. Loomis's career as a carpenter in Westford stemmed directly from the conditions of the craft there at the time. The early part of his experience indeed seems to have fit the pattern of local wandering mentioned earlier. Here was a man who either came directly to Middlefield from Massachusetts or, as seems more likely, arrived there from yet a third place. He then moved over the line into Westford—along with the other members of that extraordinary crop of 1855—to practice there, perhaps in response to a job that may have come his way through the kinship network of his wife's large family. But once his wife died in 1855 and he himself was excommunicated from the Westville church, the forces that were moving Charles Loomis seem not to have been related to his work. Of course, he might have weathered his troubles better and stayed in Westford with his daughters if he had had solid opportunities and kin of his own there. As it was, he chose to leave, "backtrailing" to a place where his family ties indeed were many. What lesson is to be drawn from what is recoverable of his story? One point is that the carpenter population of Westford included, along with stalwart native sons such as James Holmes, a fairly sizable number of men who never had the personal or family resources or the economic opportunity to make themselves fixtures in the vicinity. In his fugitive condition, Loomis represents many other outsiders who did not find in Westford the means sufficient to a permanent career.

THE CASE OF WILLIAM H. GROFF

The story of another migrant carpenter who was a sometime resident of Westford, William H. Groff (no. 23), reveals different patterns of transience. Unlike his near-contemporary Loomis, who was tied to Westford only by his in-laws, Groff had family resources of his own in Westford. Born about 1819 in Rensselaer County, he had come to Westford on his own in 1840, when he already was a young adult, evidently choosing that township as his destination because various kin were already established there. The 1840 census contains no separate listing for him in Westford. In all probability, he was the fifteen-to-twenty-year-old male living with his childless kinsman Nathaniel Groff, a native of Rensselaer County, who had migrated to Westford around 1815 with his father and mother and several of his nine siblings. Although the exact relationship between the two Groffs is not clear, Nathaniel's father, Paul Groff, probably was William's uncle. What we seem to have here is an instance of chain migration, the same pattern discernible in the history of the Holmeses.

Nathaniel Groff and his wife, Amanda Tyler, farmed in Westford, and it was as an agricultural worker that Nathaniel's cousin William was counted in the 1840 census. But the young man did not remain in agriculture in the years immediately following. By 1850 he was listed as a householder, but not a property owner, in his own right. He was married, had two children, and was living next door to Archibald Groff, one of Nathaniel's brothers. William's wife, Sarah Jane—they must have married around 1840 or 1841—was the daughter of carpenter Richardson Babcock. It is likely that William, listed as a carpenter himself, learned the craft from Babcock, with whom he may have been working. By 1855, Groff must have been working more or less on his own as a carpenter, or at least not with Babcock, who had left Westford in the interim.[35] Apparently, things were going well enough for thirty-six-year-old William. He and Sarah Jane owned land, probably the plot where their $400 frame house stood, along the banks of Little Elk Creek a couple miles south of the village.

And apparently, things remained good for the Groffs over the next few years. By 1858 William was listed in the local assessment book as the owner of one hundred acres of land worth $1,800, while in the census two years later he claimed real estate worth $3,500 and a personal estate of another $1,200. But even if he owed some of his rise in the world to his success as a carpenter, he was no longer calling himself that in 1860, when the census placed him once more in the southern part of the township. A farmer pure and simple by that time, he may have followed a trajectory similar to that being pursued in just these years by Milton Wright and by James Holmes about a decade later. That is, he may well have entered the craft largely as a means of amassing capital so that he could purchase the land necessary for farming. The 1863 assessment book shows him to have added another twenty-three acres to his holdings, the whole now worth, due to revaluation, only $1,550 total. But in the years immediately afterwards he must have altered his plans or suffered reverses. Although he still called himself a farmer in 1865, he was living next door to his cousin Nathaniel Groff and quite near James Holmes in a house in Westford village worth just $550. He probably owned that house: in the 1868 assessment book he was credited not only with a thirty-three-acre outlot worth $600 but also with a one-half-acre village lot worth $700. Now supporting four children ranging in age from five to sixteen (one born in 1842 was no longer with the family), William Groff does not seem to have prospered during the war. Nor does his scheme of switching from carpentry to agriculture—if such *was* his scheme—appear to have worked. He had managed to avoid service in the army during the past half-decade, but where exactly was he headed?

In fact he was ready to leave Westford, and like many Otsego residents in this period, he did not make a timid move. The enumerator for the state census in Otego township in 1855 noted that "near 100 inhabitants went west this spring." His counterpart in Unadilla twenty years later noted at more length, "The tide of emigration west has taken many on its flow[.] The old have passed away[,] the young have sold their birth right and mooved [*sic*] with the throng." It was to people such as William H. Groff and his family that the census-takers were referring. Groff's "west" was Minneapolis, where once again the influence of kin seems to have urged him to relocate. There, at least since 1865 (so the directory for that year reveals), one Edgar A. Groff had been in partnership with Elisha W. Eddy in a livery stable. In 1867, Edgar A. Groff was working in the lumber mill of Jones & Co., while two other Groffs—probably his sons, W. H. and H. J.—were listed as carpenters. Thereafter, the second of these does not appear again in Minneapolis directories, but the first most definitely does. The two young carpenters' father probably was the brother of William H. Groff, after whom it seems likely that W. H. was named. William H. Groff himself first appears, as a carpenter, in the Minneapolis directory in 1869, and he reappears there either as a carpenter or a millwright until 1881–1882, when his listing was changed to contractor. Young W. H. Groff, on the other hand, is absent from the listings from 1868 until 1876, when he reappears as a carpenter once more. He had the same listings for the next two years, after which he was associated with the firm in which his father once was a partner—E. W. Eddy & Son—as a floorman, a foreman, and a laborer.[36]

Although the exact details of his relationship with the Minneapolis Groffs are not clear, William H. Groff thus seems to have benefited a second time from chain migration. His move away from Westford probably brought him some new prosperity, for the millwright's trade was an exclusive and demanding one, and the expanding grain center of Minneapolis was indeed a more promising site for either a millwright or a carpenter than any location in Otsego County could be. Yet not everything went well with his family in its new home. As the Schenevus paper reported early in September of 1868, his wife, Sarah Jane, had died in Minneapolis on August 12. Her death is corroborated by an earlier story in the *Minneapolis Daily Tribune*, which reported that the cause was consumption and that the funeral was to be held at William's house at the corner of Tenth and Ames streets on August 13.[37] The family's friends were invited to attend.

Back in Westford on the day before her death, ironically enough, there died one William H. Groff—brother of Nathaniel and Archibald, usually called Harrison, probably to distinguish him from his kinsman, with whom he shared the same birth year as well as name. His Minneapolis namesake, having drawn on a wide kinship network in Westford, could not at last succeed there. Only when he went west, as so many Otsego natives themselves were doing, did he find what seems to have been an adequate niche for himself. He was merely one of many craftsmen spawned by Westford or at least hosted by the township who, as its population fell, could not successfully ply their craft there. Such were the men whose absence in 1869 made that year so full of possibility and accomplishment for James Holmes.

Conclusion: Skill and Responsibility

William H. Groff's career is in several ways the perfect counterpoint of Holmes's. Holmes preserved his ties with the township at the expense of his craft, whereas Groff remained active in the craft only by leaving behind his adopted home, much as he earlier had left behind his birthplace and come on to Westford. Such were, in large part, the choices that these workers faced.

Remaining behind was indeed a *choice*. It would be wrong to think of James Holmes or the few others who stayed in Westford as merely the victims of an economic dislocation. As I noted earlier, for one thing, Holmes's shift from carpentry to farming in the 1870s—coinciding as it surely did with the depression of those years—may have been part of a positive strategy designed to fund his farm by the paid labor of his hands. Once he had purchased land on the Worcester road and built his farmstead there, he was ready to begin the transition. For such a man, clearly enough, moving to Minneapolis—or even Oneonta—hardly would make sense. It was not the city he wanted, or not the economic opportunity the city offered and represented. It was a more traditional economic life, one ensconced in the land and therefore available best of all—if he could get a start there—where his kinship and social networks could help him. Even when he began farming he pursued not the market but a kind of self-sufficient diversity.

It would be unfortunate, at the same time, to cast our more rigid notions of occupation back onto Holmes's world. If he did change his declared occupation in the 1870s, he did not thereby cease to be a member of the craft that had funded his purchase of the farm he was now more seriously tending. Carpentry was as much a skill, a kind of knowledge, as a gainful employment. It could also be the latter, but within the lingering premodernism of rural America in the nineteenth century, such skills were more nearly a matter of culture than economics. (But, as an aside, it seems worth pointing out that economics itself is always tied to culture: the economic assumptions that would urge us to divide Holmes's life otherwise are, after all, part and parcel of modernism itself.) The resurgence of activity among other carpenters than Holmes after long periods of apparent inactivity suggests precisely how imprecise the boundaries of carpentry were in such places as Westford. Milton Wright thus shows up as a farmer in 1860 after two census listings as a carpenter, but by 1865 once more he is a carpenter; William Bentley, active in the craft locally from as early as 1814 until the mid-1840s, described himself as a farmer in 1850 and 1855 but then declared in 1860 that he was a carpenter; Hezekiah Skinner, a nineteen-year-old carpenter in Delaware County in 1850, still described himself as one in Westford five years later, then was a farmer in 1860, only to call himself a carpenter again in nearby Worcester in 1865. To be a carpenter, especially in this world of composite economic life, was to acquire certain abilities that hardly disappeared merely because one made much of one's living by other means. If nothing else, carpentry could provide a farmer with a handy source of day hire or short-term work during slack agricultural periods or when ready cash was needed for some project on the farm. Since few of these men, even when declaring themselves carpenters census-by-census, seem to have had steady craft jobs—that was, after all, an urban-industrial phenomenon—their passage into and out of activity was not the epochal event that we may think it. This was not a matter of mobility so much as flexibility.

It would be wrong to conclude, too, that the apparently high mobility of Westford's carpenters meant that this was largely an alienated workforce connected by merely casual ties. To the contrary, it is clear that there were certain strong kinship and social patterns among these men and that a few relatively durable individuals served to give the local craft a rather impressive solidity. To begin with, it is possible to show that a majority of the practitioners were associated with local masters who either trained or employed them, and this shop tie was often doubled or tripled by kinship or associational links or both. Milton Wright, William Bentley, and Andrew and Jacob Snyder, to name four of these masters, were responsible for training a large number of other practitioners. Milton Wright thus raised his son William and his nephew Judson Winans, both of whom were carpenters, and he probably helped train his two half-brothers, Henry and Warren Wright, as well. In addition, Holmes's brother John and Holmes himself both were part of Milton Wright's craft circle. The Bentley circle included William's nephews Edgar and Stephen as well as Ezra Thurber and Ezra's son Oliver. Andrew and Jacob Snyder, who were brothers, were at the center of a group that included Jacob's son Chancellor as well as Erasmus Snyder, the son of Andrew's twin brother, Christopher. In the period from 1865 on, it was the Holmes brothers and their kin who constituted the last important craft network in Westford.

Kinship and associational ties were crucial in such networks. Many of the carpenters most closely linked to Westford were related to two or more of the master groups, often via social and kinship bonds that solidified the more nearly occupational ones. Oliver Thurber thus was most closely related to his father, who had worked for William Bentley. But when Oliver married Abigail Snyder McClintock in the 1870s, following the death of his first wife, he allied himself with the Snyder clan as well. Similarly, James Holmes was linked to Milton Wright not only through his own probable training at Wright's hands but also by the fact that his brother John had previously been trained by Wright. Furthermore, Milton Wright and James Holmes were both Baptists centrally involved in the Westford congregation, which each served as clerk (Holmes, it will be recalled, was elected to replace Wright early in the 1860s) and which together they served as members of the committee charged with overseeing building renovations in 1860. Additionally, John Holmes had been in business with Milton Wright's half-brother Henry before the Civil War, while Milton's son William was James and John Holmes's brother-in-law (and was employed by James briefly during 1869); John Holmes married Milton Wright's half-sister Louisa in 1851, and James Holmes not only was employed building Jud Winans's house but in turn employed Winans on several other projects. Such interlocking ties no doubt helped determine which among Westford's too-abundant carpenters remained in the township and had a chance to work at their craft there. As the available jobs decreased, it appears to have been the more stable circles that tended to get them. Certainly this was true of James Holmes as the beneficiary of Milton Wright's influence in the late 1860s. Probably the same concentration of power accounted for the fact that so many outsiders who entered Westford in search of craft work seemed unable—even in such a case as William H. Groff, who after all had some kinship ties there—to find an opening for themselves locally. Hence the fugitive, at times even mysterious, nature of their swift passage through the surviving records.

On the other hand, the fact that Milton Wright himself and his son William and nephew Judson Winans left Westford in the late 1860s suggests how slim were the chances there even for the old guard by that time. But that James Holmes and his brother John did *not* leave then may be owing to the importance of ties other than occupational. To be sure, they had the most stable positions in the village, since they were the two carpenters with the longest number of years in residence in Westford. By 1880, however, neither was declaring himself a carpenter anymore: James diverted his energies to his farm, while John worked at wagonmaking and then was simply a day laborer, the lowest possible job that a man as skilled as he was ought to have had. Had the Holmeses been willing to relocate, the Wright-Winans connection probably would have stood them in good stead. In addition to Milton and William's carpenter work in Schenevus and Judson's in Oneonta, there was the fact that Milton's two daughters had married the owners of the Schenevus sash-and-blind factory, a business that employed carpenters (of a sort) and would have been a logical clearinghouse for carpentry jobs in the vicinity.

That the Holmeses did not avail themselves of such opportunities is clear; exactly why they did not isn't. Part of the reason may have been that they were firmly attached to Westford. With the exception of James's years of medical training, they never really left the township. By contrast, the Wrights and especially young Winans—not to mention several of John and James Holmes's own younger siblings—were less tightly tied to the area. Perhaps the fact that Milton Wright had left

Westford earlier in his life, working in Worcester at his craft, suggests a longstanding, lower degree of attachment in his case. Numerous of his comments in the Baptist clerk's book derogate the congregation and at times lead up to halfhearted resignations of his post; they likewise suggest a man of uneasy, though intense, affiliations. Milton's son Will had his own brief time away, as a young soldier in the Union Army, an experience that must have weakened his attachment to Westford somewhat and may have contributed to his move to Schenevus. But Will Wright *did* return to Westford, for longish though intermittent periods, and he ultimately died there. By contrast, the equally brief army experience of Will's uncle Warren, a Westford sash-and-blind maker before the war, seems to have ushered in a final break with the township. Warren probably never even lived there after he was discharged, as on the contrary young Will at least did before leaving again in the later 1860s. Will's other kinsman, Jud Winans, transplanted several times and orphaned in Michigan, also had local ties much weaker than Will's. Westford probably was the home of Winans's parents, but his own residence there after 1855 had been an act of necessity rather than choice. Although he continued to live within Otsego County until his death in the late 1890s or the early twentieth century, once he left Westford in 1870 he was gone from there for good.

James Holmes, by contrast, made his home there along the Worcester road until that Sunday morning in March 1895 when, after church, he went across the way to tend his cows and then died in the barn that, with the help of Jud Winans, he had built in 1869. We may be tempted to see him nostalgically during the last decades of his life: as a carpenter whose "useless arm" (to borrow a phrase from his diary) was the mark of his economic marginalization by the increasingly urban, mechanized world to which Winans migrated when he moved to Oneonta. But Holmes's fate was in part the result of positive choices he had made. To be sure, little that he did while plying his craft was unaffected by modernization. For his "house for self," he indeed had produced a heavy timber frame hewn from hemlock cut from a stand so close he could see it from his new front doorway. And the stone for the foundation came from a quarry in which, as a boy, Holmes must have played. It was laid up, moreover, by a neighbor whose house in turn Holmes was to erect, on a lot a few hundred feet away, later in the year. But there the localism ended and the modern cultural order took over. The plan of the house, a simplified copy of the Greek Revival structures that even little Westford boasted, derived from a national love affair with things classical. That was a love affair that sought to express the country's faith in political immediacy, cultural innovation, and native talent. Given the association of Greek Revival buildings with such values, it is ironic that beneath their white-painted wood-clad American exterior they bespoke things distant, old, and foreign. Ironic, too, was the fact that the finer elements in the Holmes house, things that, like the sash and doors, would have been made locally not many years before, came now from beyond the immediate community. The steam-powered mills where those items were produced were stripping independence, pride, and hope from craftsmen like Holmes. With all the limitations of their world, such workers once controlled far more of their labor and their lives than their children would be likely to.

Still, we can read Holmes's refusal to follow his craft wherever it might lead, and to abandon his home and family in search of the main chance, as an affirmation of what mattered most about the old ways. Here, in the upland isolation of southeast Otsego County, Holmes may have found a modest installment of that rootedness that his forebears, forsaking any number of homes, had sought to discover in landscape after landscape. As his neighbors died or moved away and his livelihood shrunk, the land, with its wonderful upcast of hills, continued to give to those who sought it with patient grace something beyond the success that cost so much, whether it lasted or not. In the photograph that Holmes and his second wife, Daphnea, sat for with their son Ernest in 1895, the year of Holmes's death, there is no mark of defeat or bitterness on his placid face (fig. 46). With his heavily veined workman's hands closed gently on each other and his legs angled out with an ease that belies his age, he sits upright and looks directly at the lens. On his face we do not find the downturned mouth of Daphnea, who seems more resigned than her husband. No, for him there is instead a kind of acceptance. He lacks the jaunty energy that young Ernest projects, madly dressed as he is for the new age about to come. But Holmes shows no real weariness either. He takes possession of the

46. James C. Holmes, his second wife, Daphnea V. Cook Bentley, and their son, Ernest, 1895

frame as firmly as he must have taken hold of those acres that remained his family's home to the end. It was a modest life but a good one: a life informed by skill and as true-tempered as a good steel plane. And neither of his arms looks "useless."

"Skill, in the best sense," writes Wendell Berry wisely, "is the enactment or the acknowledgement or the signature of responsibility to other lives; it is the practical understanding of value. Its opposite is not merely unskillfulness, but ignorance of sources, dependencies, relationships."[1] Though he was no philosopher and surely would have had some difficulty understanding the terms of Berry's wisdom, James Holmes did not define himself unidimensionally, as the modernizing revolution of his time urged people to. He was not *merely* a carpenter. Holmes in effect lived in accord with Berry's precept. If the skill with which he practiced his craft entailed a kind of acknowledgment, so also did the act of staying put. Doing so depended on two related recognitions: accepting his life as part of a larger social and environmental whole and refusing to accept any part of that life as itself the whole.

Even the way in which Holmes practiced farming suggests the centrality of such values in his life. I have not tried to tell the whole story of his life as a farmer here, but I want to say just a bit more on the subject. As noted in the first chapter, his farming resembled his father's traditional practice much more than it did the modernized, intensive agriculture of his brothers Joseph and Horace. We would be wrong, that is, to think that Holmes abandoned carpentry for farming simply because the former did not assure him success and the latter did. He was not an aggressive farmer, one who aimed at making a profit from the land before all else. Indeed, land records might be read as suggesting that he failed at farming. Ironically, his urbanite sister Clarissa wound up owning Holmes's farmhouse—the one he'd built for his family in 1869—probably because loans she was making to Holmes even in the later 1860s were never repaid. Only when Holmes married Daphnea Bentley in 1881 was Clarissa paid off, and it was his new wife who discharged the debts, thereby becoming the legal owner of the house. Daphnea also bought Holmes's farmland from him in the same year, so that when he died in 1895 neither the land nor the house belonged to him any longer. Although these transfers of title may not have meant much in practice, they certainly suggest that Holmes's farm hardly was a source of much cash. But such a state of affairs should not necessarily be glossed as a failure. Holmes seems not to have had any real intention of engaging the agricultural market, and hence his farm never could have provided him with the sort of income that his brothers must have enjoyed from their more specialized and larger-scale operations. Instead, Holmes raised some of his family's food on the small acreage that he called home and sold surpluses, much as American farmers in such inland areas had been doing for two centuries.[2] He seems to have understood well the difference between living and making a living. If there was an average Westford farmer in these years, especially after the railroad made the link with urban markets more immediate, that figure must have existed somewhere between the two extremes represented by James Holmes and his brother Joseph. The more affected by—or, we might say, converted to—modernization, the closer any farmer would have been to Joseph's position.

And the future lay largely with the modernist party. The choices of Holmes's daughter Nettie, very different from his own, illustrate the divergence here, although with an ironic twist that reminds one of the pyrrhic victory that modern farming has often visited on the American farmer. Nettie had kept house for Holmes following her mother's death. Then, when he remarried, she left Westford, going north into the heavily industrialized Mohawk valley. There she must have worked in one of the gray, stonewalled factories that turned out leather goods or textiles. And there, too, she met George Shattuck, whom she married in 1882. The young couple apparently wanted to leave mill labor behind and go back to the land. Maybe they looked for an acreage in the immediate vicinity; perhaps they spent a brief period back in her native village seeing what might be available there. But soon they turned their eyes—as so many local inhabitants had—toward the beckoning West. Responding to the propaganda that urged would-be farmers to settle in the semiarid high country of western Kansas, they took their young son, James, and moved in 1886 to the region surrounding Bird City. There, almost on the Colorado border, they dug the turf and made a house from it. It was a modest shelter, but its sod-covered roof, when the rains came the following spring, bloomed with the still unconquered prairie flowers. It was a sight that their son Jim never tired of

47. Nettie Holmes Shattuck (left) and her family, ca. 1920

recalling. He also told tales of how the snows had come thick in the winter, driven by a wind Otsego never knew. And of how the drought of 1894–1895 had burned them out, sending back to the East not only the Shattucks but also thousands of other families that had crowded this most unlikely agrarian paradise over the past decade and a half.

They had all believed Charles Dana Wilber, the Nebraska land speculator who in 1881 had coined the ringing phrase: "Rain follows the plow."[3] They had believed him too literally. The storms of the mid-nineties, storms that were the ultimate bloom of their own plows, showered them with dust instead. The Shattucks got back to Otsego eventually, having lost all that they had built in the nine years when Jim was growing up. From then on it would be a story of rented farms around Westford and then further west in New York (fig. 47). Jim stayed on in Otsego, renting farms with his wife, Pearl, until their herd of cows was condemned as tubercular and shot. After that, when he had become a widower himself, he moved to the Mohawk valley much as his mother had in her youth. Marrying again—his new wife, Flossie, was kin, his sister-in-law, widow of his first wife's brother—he took a modest job as a church sexton.

Some things remained.

48. James C. Holmes's secretary desk, ca. 1860

On his death in 1968, there stood in Flossie's small house a secretary desk that had held special meaning for him. It was a country piece (fig. 48), with boldly turned legs and a severe though beautiful look, small inlays of curly maple setting off the pine and cherry panels. When his parents moved west in the 1880s, they had sold off this piece along with other furniture. And after Jim was back in Otsego, the young man searched the desk out in country auctions, finally locating it and buying it back. Why was he so diligent in doing so? It must have been a piece that his grandfather, the carpenter for whom young James Holmes Shattuck was named, had made years before there in Westford. That Holmes was a cabinetmaker of sorts as well as a house carpenter the diary of 1869 indicates—it records, after all, Holmes's construction of a wood wardrobe for his uncle, James Kelso, early that year. Such a mixture of skills may have been another means of surviving in this difficult locale, but it also was another mark of old ways that Holmes carried forward: Westford carpenter William Bentley, active early in the century, had also made chests and wardrobes and wagons, as well as houses and coffins, for his neighbors.[4] Whatever was uncontained in a neighbor's life the local carpenter could provide the wooden case for. Appropriately, Holmes's grandson kept the small pocket diary for the year 1869 in one of the maple-veneered drawers of the desk, both text and container becoming tokens of the carpenter's now silent skills, a kind of condensed "home" young Jim could carry along with him to represent the old center of the family's life. Beyond the enormous changes his own times had witnessed, a few tangible things tarried from the past. However much he had been displaced from the landscape of his familiar history, memories could not be dislodged from his heart.

The stories he told me as a boy were to lead me on a search, though I never imagined so then. It was a search that helped me to reimagine the world of Jim's grandfather, to reimagine it and to recreate it here.

General Note on Sources

Information on individuals discussed in this book has been derived from a number of published and manuscript sources. Rather than indicate in each instance the source of any assertion, I shall indicate here the nature of the materials relied on. In most instances, the context makes the specific source clear; where this is not the case and the issue in hand is important, I have made direct citations, as I have when directly quoting any source. Materials used only for identifying a single individual of course are cited fully at the appropriate point.

Information regarding the dates for individuals derives from three major sources. Various cemetery transcriptions from Westford, Maryland, Worcester, Middlefield, and Decatur town burial grounds, while not infallible or complete, offer the fullest guidance. These transcriptions are housed at the New York State Historical Association (NYSHA) in Cooperstown. United States Census and New York State Census population schedules for the years from 1850 to 1900 (housed at the county clerk's office in Cooperstown or used in microfilm versions) provide inferential (and in the case of the 1900 federal schedules, more exact) information regarding birthdates, although comparison of figures for given individuals over time suggests that these be used with some skepticism; in a few cases census mortality tables also provide information regarding dates (and causes) of death. Finally, a variety of local materials has been consulted. Foremost among these was *A Century and a Half in Westford*, a mimeographed volume prepared in 1938 by students at the Westford Union School. The advisor for this volume, Lewis J. Holmes, for many years Westford's town historian, was gracious enough to meet with me twice while I was at work on this study. His extensive knowledge of his hometown was a great aid; the fact that the carpenter-diarist James C. Holmes was his great-great-uncle gave his knowledge a special poignancy for me. Lewis Holmes has helped many other people—like the students of 1938—understand Westford's past. Among his beneficiaries was Raymond C. Tyler, a Westford native, who in the 1960s prepared a manuscript "History of Westford," a very well illustrated volume (housed at NYSHA) that has been the second most important local source for my work. In addition, such materials as newspaper obituaries, the Westford Baptist church clerk's books (Baptist Book A and Baptist Book B, both housed at NYSHA), family bibles (at NYSHA), manuscript diaries, and the like have helped me fill in a few gaps. I made some use of biographical and other information contained in the *Biographical Review . . . Biographical Sketches of the Leading Citizens of Otsego County, New York* (Boston: Biographical Review Publishing Co., 1893; henceforth, BR.93) and D. Hamilton Hurd's *History of Otsego County, New York, with Illustrations and Biographical Sketches of Some of its Prominent Men and Pioneers* (Philadelphia: Everts & Fariss, 1878), as well as (for a number of ministers) *The First Half Century of Madison [Colgate] University, 1819–1869* (New York: Sheldon & Co., 1872; henceforth, FHC.72). I have given firm dates for individuals only when I have been satisfied that the source or sources I have used could be trusted. Otherwise, I have given approximate dates. It is almost always true that approximate dates have been extrapolated from the census schedules and firm ones derived from cemetery transcriptions or local records.

Information regarding occupation and residence in Westford (or elsewhere) at a given time usually derives from the census schedules already mentioned. When the year in question is either the first or the sixth of a decade (e.g., 1850 or 1855, etc.), it may be assumed that the source was, respectively, the U.S. or state census for the given year. In addition to the local materials referred to above, I have also used three other important sources of information regarding occupation and residence. The first is a group of local directories dating from the decade following the end of the Civil War, including one for Otsego County, two for the state, and two for the line of the new A & S Railroad.

Hamilton Child, comp., *Gazetteer and Business Directory of Otsego County, N.Y. for 1872–3* (Syracuse: The Journal Office, 1872), henceforth, OTCO.72.

Sampson, Davenport, & Co., *The New York State Business Directory, 1870, Containing the Names, Business, and Address of all Merchants, Manufacturers and Professional Men Throughout the State* (Albany: Charles Van Benthuysen & Sons, 1870).

Andrew Boyd, *Boyd's New York State Directory. 1872–1874* (Syracuse: Truair, Smith & Co., 1872).

Andrew Boyd, *Business Directory and Gazetteer of the Towns and Villages on the Albany & Susquehanna Railroad; and of Cooperstown, Delhi, Norwich, Oxford, Franklin, etc., with Much Interesting Information, 1868–9* (Albany: Charles Van Benthuysen and Sons, 1868).

Andrew Boyd, comp., *Boyd's Binghamton City and Susquehanna Railroad Directory* (Binghamton: H. E. Pratt, 1871).

The coverage in these various directories is never so inclusive that anyone's absence from a listing can be taken to mean very much; but his or her presence, especially when it is confirmed from other sources, usually may be assumed to be accurate. A second, similar source, is the *Atlas of Otsego County* prepared under the direction of F. W. Beers and published by him, A. D. Ellis, and G. G. Soule in New York in 1868. Its town maps and village plans contain a great deal of detailed (but not always accurate) information regarding inhabitants and their occupations. A similar but far less detailed wall map of the county published by C. & B. C. Gates in 1856 also has been consulted. Finally, there is a series of manuscript tax-assessment books surviving for Westford (for the years 1858, 1862, 1863, 1866, 1867, 1875, 1876, 1877, 1878, 1879; now in NYSHA); although such records cannot in themselves confirm someone's residence in a given area, they do provide valuable hints on that topic, as well as firm data on land ownership, and some vague indication of current occupation (craftspeople and merchants, for instance, often owned only a so-called village lot or at most a small farm).

Information on mobility has been gathered primarily from census schedules for surrounding towns, other counties in New York, and (on a couple occasions) other states; but some use has been made of the Westford Baptist church records (and those of a couple other nearby congregations, also in NYSHA), since these on occasion indicate when a member left or arrived in the town and where members came from or went. Data on farm size and farming activity have come from the assessment books mentioned earlier, as well as from the agricultural censuses taken by both the federal and state governments in conjunction with their stated population surveys, and from the 1872 Otsego County directory already referred to. I have also made some use of published reports of both the federal and state censuses, usually for the purpose of placing Westford figures in context.

Appendix 1

Westford in 1869: A Biographical Guide

This guide is divided into two parts, the first covering the immediate family of James Holmes and the second covering other people (both kin and non-kin) mentioned in the diary. In part 2, italicized names at the head of an entry give the form when it is confusing or differs significantly from that used in the guide, in which Holmes refers to the individuals in question in his diary.

1. THE HOLMES FAMILY

James Holmes came from a large family, and in 1869 he was building his own smaller one.

JAMES HOLMES'S PARENTS

Calvin Holmes (1796–1869), James's father, was a native of Salem, Connecticut. He had moved to Westford around 1818 at the urging of his brother John P. Holmes, apparently for the purpose of teaching in one of the new local schools. Calvin's brother, who had emigrated to Westford in 1812 but later moved his own family to Cortland County, himself had taught in Westford, but Calvin seems not to have. Soon after his marriage in 1819, he began his long career as a farmer on Westford's hilly terrain. In late February 1869, he took sick with a "bilious fever" and, improving and then worsening over the next three weeks, ultimately succumbed to the disease. James Holmes sat up with his father all night on three occasions, "washed and dressed" him (3/7 entry), and toward the end visited him every day.

Mary Kelso Holmes (1800–1845) married Calvin Holmes in Westford in 1819. Mary Kelso was the only daughter of a Massachusetts-born blacksmith; at least two of her brothers practiced the trade in Westford as well. Mary Kelso Holmes had twelve children before she died in 1845.

THEIR CHILDREN

Mary Lucy Holmes (called *Lucy* in the diary; 1820–1885) married the farmer William T. Marks, a native of Cherry Valley, in 1849. They lived there for a time and then moved to nearby Clarksville in 1850 before eventually coming back to Westford. James Holmes had several dealings with her husband in 1869.

Fanny Almira Holmes (1824–1904) in 1844 likewise married a local man, James B. Skinner, and set out farming with him in the area called Skinnerville north of Westford village. The Skinners had important dealings with Holmes in 1869: they loaned him $100 in September, and in October one of Holmes's workers did some sizable unspecified job for them. In the 1875 state census, Fanny A. Skinner was described as insane, a designation that may have been inaccurate, however, because it was absent from all her other census listings.

Joseph Marvin Holmes (or Marvin Joseph; 1826–1894) married Perlina Washburn in 1846 and thereafter farmed large acreages northeast of the village. In 1869, James Holmes hired Joseph's son Revilo to teach in Westford District School No. 3, of which James recently had been elected trustee.

John Kelso Holmes (1827–1907) in 1851 married Louisa A. Wright, the half-sister of the carpenter Milton Wright, who almost certainly trained him in the craft. John Holmes spent his whole life in Westford but passed through a series of occupations unlike those of James: from carpenter to sash-and-blind maker to wagonmaker until (in the 1900 census) he had become a simple day laborer. He did not work with James in 1869, but it is likely that they labored together on various occasions.

James Calvin Holmes (1830–1895).

Horace William Holmes (1832–1910), the sibling closest to James in age, spent his life in Westford. He farmed with his father, Calvin, living with him and then next door to him until Calvin died in 1869. He and his wife Lucinda Skinner (sister of James B. Skinner; they married in 1856) then bought out Calvin's other heirs and farmed that land until their deaths. During 1869, James had many dealings with Horace: Horace loaned him money (memorandum 2), sold him hogs and two cows and mutton, and even put in some time on James's new house.

Clarissa (or *Clara*) *Elizabeth Holmes* (1833–1920) did not stay at home, although her visits to Westford in 1869 (described in the diary) suggest she hardly was out of reach. A dressmaker in Westford in 1855, she left for Albany in 1857. Later she was living and apparently working in the town of Coeymans south of that city, then in Albany proper, perhaps in her old trade or as a schoolteacher. That she was doing well is suggested by her many loans to her brother James (and to her brother Horace) both in this period ($50, for instance, in January 1869; memorandum 2) and in later years. In 1891 she was to marry the patent-owner and businessman David W. Seeley. It was with Seeley and his first wife, Maria, that James Holmes was staying in Albany when he began his 1869 diary.

Julia Sophia Holmes (1836–1908) was less reachable than Clarissa. In Westford in 1857 she married Alpheus B. Chapman, with whom Mary Shelland and then Mary and her husband, James Holmes, had shared quarters in the house of Chester Wright earlier in that decade. Soon after Julia and Alpheus married they moved to Illinois, where they spent the rest of their lives. In his diary, James notes telegraphing them about the death of Calvin in March, only to find that they couldn't come for the funeral. They did manage, however, to visit Westford during the Christmas season.

Emily Jane (or *Jennie*; 1839–post 1931) taught school in Westford for several years prior to 1860, when she joined the Chapmans in Illinois. There she continued teaching even after she married, in 1862, Absolum O. Edison, reportedly a cousin of Thomas Alva Edison.

Sarah Amelia (1842–1905), the youngest of James's siblings, married the carpenter William E. Wright, son of Milton Wright, in Westford in 1867. They soon after moved to the nearby village of Schenevus but remained there only a few years before returning to Westford, where they were to die. James worked with Will Wright on at least one job in 1869.

(Mary Kelso and Calvin Holmes also had two children who died young: Lyman Leander [1822–1850], who seems to have left the area in 1845, and David Henry [June–September 1838]).

JAMES HOLMES'S OWN FAMILY

James Holmes, once he returned from his stint as a student of medicine in nearby Delaware County, began work in Westford as a carpenter. In May 1855, in the Westford Baptist parsonage, he was married.

Mary Shelland (Shellan/Sheldon) Holmes (1835–1878), his wife, was the daughter of David (ca. 1795–1869) and Marinda Shelland (b. ca. 1800). She was born in Schenevus, but by 1850 she and her family had moved to Westford. David, who was working there as a blacksmith, was living together with Marinda and their elder daughter Orilla, while Mary, just fifteen, was living (apparently as a hired girl) with the family of Chester and Irene Wright. More recently, the Shellands had been living in Worcester. David died in March 1869, as the Holmes diary records, and by year's end his widow had moved in with Mary and James Holmes at their new house in Westford.

During 1869, the Holmes diary offers some insight into Mary's typical activities. Probably her most important economic contribution to the family was her production of butter, which brought a good deal of cash into the household. But Holmes himself seems to have handled the money thereby produced, and on a few occasions (as on March 24, when he notes that he gave her $1.50 to buy a dress) he doled out cash to her for her expenses. As a housekeeper, she obviously worked much of the time also contributing to the family's welfare. And she went beyond the normal in June, when she "painted [the] chamber," and in August, when she picked hops at a nearby farm.

The diary shows her visiting Holmes's kin (his brothers Horace and Marvin, his sister Lucy), attending church, taking their daughter, Nettie, to one of the Bolles lectures in November, and even sitting by the bed of James's dying father in March.

Mary Shelland Holmes was the mother of the four children listed below. She died in July 1878 from what a local paper described as an "affection of the liver" that had bothered her for a number of years (*Worcester Times*, 7/12/1878).

CHILDREN OF JAMES AND MARY HOLMES

Arthur C. Holmes, the oldest, had been born in 1857 and was twelve in 1869. Apparently dead by the 1920s, he had been listed in the 1880 Westford census as a laborer but has proved untraceable beyond that point. His name does not appear in the records of any cemetery in the Westford vicinity, and he presumably left the area. The 1869 diary shows him as a choreboy around the Holmes household. Most importantly, he helped his father cut, draw, and stack firewood, and on his own he drew and spread manure from the family's livestock. He also drove Holmes's team around on errands (drawing shingles, wood, stone, and corn) and ferried various visitors about. His importance to Holmes is suggested by the March 25 entry: "Arthur [was] sick. I had the chores to do."

Nettie F. (also called *Fannie*) *Holmes*, born in 1861, was eight in 1869 and is mentioned only a couple times in the diary, once when Holmes records having bought her some shoes and once when she attended one of the Bolles lectures with Mary. After Mary's death nine years later, Nettie began keeping house for her widowed father (so the 1880 census reports), but she must have left for Hagaman's Mills near Amsterdam just after or perhaps before his remarriage in mid-1881. Nettie married George F. Shattuck (b. 1853) in her new home in September 1882 and had a son there in 1883, but she spent the years from 1886 to 1895 homesteading (without much luck) in far western Kansas. She then returned to Otsego County for a number of years before finally locating in Springwater, south of Rochester. She died sometime in the 1920s at a son's house in Geneseo.

Calvin James Holmes, born in 1865, was four in 1869. Still in school at the time of the 1880 Westford census, he was gone by the 1900 listing. At the time of Nettie's death in the 1920s, he was living in Canajoharie, in the Mohawk valley.

John K. Holmes, born in 1871, was at school in 1880, but by the time of the state census in 1892 he had left home. In the 1920s, when his sister, Nettie, died, he was living in Santa Barbara, California.

HOLMES'S SECOND FAMILY

Three years after Mary Holmes's death in 1878, James Holmes remarried.

Daphnea V. Cook Bentley (1843–1917), his second wife, was the widow of Westford carpenter Stephen Bentley, who had died at the age of thirty-six in 1880. James mentions Stephen Bentley once in his 1869 diary (on February 26, as the Cash Account makes clear), so although Bentley is not listed in the 1870 Westford census he must have been living nearby or perhaps visiting relatives. Daphnea V. Cook was living in 1870, the census then reports, in the town of Laurens, where she was teaching school. It seems likely that she first came to Westford as a teacher, and there met and married Bentley. They were both listed in Westford in 1875 and 1880. They had no children, but shortly after her remarriage in 1881, Daphnea and James were to adopt a son, Ernest C. Holmes, who was born in 1883 (fig. 46). Following the death of James Holmes, Daphnea and Ernest continued to live in Westford until past 1900. She, however, later remarried and moved to DeRuyter in Madison County. When Ernest's stepsister, Nettie, died in the 1920s, Ernest was living in Waterloo, New York, near Nettie's old home in Springwater.

OTHER KIN

Anna Howe Holmes (b. ca. 1798) was a widow whom Calvin Holmes married sometime between 1850 and 1855. According to the 1865 New York State Census, she had three children by her first husband, who probably was the Westford pioneer Artemas Howe (1776–1849). Following Calvin Holmes's death in 1869, as the diary makes apparent, Anna went to stay with what must have been the family of her daughter in Decatur; after this, there is no further record of her in Westford and it is likely that she left the vicinity in May when Leander Roberts came from Illinois "after [his] mother."

Betsey Holmes (1793–1878) was James's maiden aunt; she was at this time the sole surviving sister of Calvin Holmes. She had moved to Westford between 1860 and 1865 from Connecticut and was living with James Holmes's sister and her husband, Fanny and James Skinner, in 1870.

Revilo Holmes (1847–1929), the son of Holmes's brother Joseph Marvin and his wife Perlina Washburn, was hired by Holmes late in 1869 to teach in the District 3 School for the upcoming winter term, which was to end the following February 25. His pay, equal to his uncle's as a carpenter, was the highest of any teacher in the town in 1869 (see the trustees' reports for 1870, preserved at NYSHA). The phrase *and board [by] self* (see 11/1 entry in the diary) means that Revilo was expected to supply his own meals rather than board with families in the district or with his uncle James (by *self* in the diary, Holmes typically refers to himself, but his trustee's report confirms the interpretation just given). Trustees were forbidden to hire anyone as a teacher who fell "in the second degree of relationship by blood or marriage" to themselves (V. M. Rice, Superintendent of Public Instruction, *Laws of New York Relating to Common Schools, with Comments and Instructions, and a Digest of Decisions* [Albany: Weed, Parsons and Co., 1868], p. 142). According to the principles enunciated by Rice, Revilo and James in fact did not fall within the proscribed limits (see p. 144). The name *Revilo*, by the way, happens to be Oliver spelled backward; but according to Revilo's grandson Lewis J. Holmes (interviews, October, 1983), the giving of this name to Revilo Holmes was in no way a reference to Oliver Wendell Holmes. Within Westford there was another Revilo (Revilo Seward; aged nine in the 1855 census and thus roughly Revilo Holmes's contemporary), and this name seems to have been merely one of several fanciful or at the least poetical male names ending in *-ilo* (again in Westford, and all in the 1870 census, one finds the twins Lavilo and Milo Goodenough and the unfortunate youngster Devilo Bulson!).

2. WESTFORD AND AREA RESIDENTS

Note: The letters *WN* preceding someone's dates indicate that he or she was a Westford native.

Michael P. Agan (ca. 1842–1888), originally from South Valley in the adjoining town of Roseboom, was married in 1866 to Mary Louisa Wright (1844–1921), the daughter of carpenter Milton Wright and sister of William E. Wright, Holmes's brother-in-law (see *Westford Literary Institute Herald*, 3/17/1866). The 1870 census lists him as a farmer.

Daniel Aldrich (WN, b. ca. 1816) was listed in the 1870 Westford census as a retired farmer. He lived in the north end of the village; in April 1868 he bought from James and Mary Holmes for $2,150 a parcel of twenty-six acres there, but he resold it to them in March 1869 for $2,000.

Daniel Nash Allen (1807–1869), who died of consumption in 1869, had farmed in Westford since 1842. Probably because of bad health, he had auctioned his belongings at his farm on Darling Hill, south of Westford village, late the previous November. His farm proper was sold to Harrison Groff. *Schenevus Valley News*, 11/28/1868 and 2/16/1869.

Fannie Allen (b. ca. 1821) became, sometime after 1865, the third wife of Daniel Nash Allen, whose death Holmes noted on February 4. In 1870 she was living in Westford with four stepchildren.

Carlos Ashley (b. 1840), living with his father in 1870, was listed as a stonemason; the same was true thirty years later.

William O. Ashley (b. 1815) arrived in Westford shortly after 1850. A stonemason, who along with his son Carlos was to do a good deal of work for Holmes in 1869, Ashley in turn hired Holmes to build his own new house. Still listed at the age of eighty-four in the 1900 federal census, Ashley was reported by the Schenevus paper to be in poor health in April and May 1901 (*Schenevus Free Press and Davenport Standard*, 4/17/1901 and 5/22/1901), but whenever he died he apparently was not buried in any Westford cemetery.

Lester Babcock (WN, 1800–1876) was a retired farmer in 1870, living in the village of Westford.

Augustus Baldwin (and his twin sister, Augusta) had been born to German Baldwin and his wife, Sabrina, in May 1846. Not resident in Westford proper in 1870, he obviously was living close enough the year before to work as a day laborer on some of Holmes's projects.

German Baldwin (1818–1882), usually listed as a farmer in the various census schedules, was a laborer in 1850. His work for Holmes in 1869 shows that he continued to perform such manual labor long after 1850. Holmes's workers carried forward a major building project for Baldwin in 1869; Baldwin and Holmes also were associated by virtue of their common membership in the Baptist church.

Nathaniel P. Barnard (1816–1889), who with his wife, Susan, was farming north of the village in 1870, was also listed as a watch- and jewelry-repairer in OTCO.72.

Albert Bell (1844–1916), a native of Oneida County, was the son of shoemaker Hezekiah Bell; he does not show up in the 1870 Westford census, but like his older brother John, he seems to have been practicing his father's craft there in this period.

Hezekiah Bell (1812–1887), a native of Oneida County who moved to Westford around 1850, practiced as a shoemaker there for a time, but by 1870 he and his wife, Sabrina, were farming.

William H. Bennett (1816–1879) was a merchant tailor in Schenevus. In 1869 he had just enlarged his store and built a new house.

Dolly Bentley was almost surely Phoebe D. Bentley (b. ca. 1807), the third wife of Baptist Deacon Russell Bentley and a member of the Westford congregation since 1845. Their son Stephen later married Daphnea V. Cook (subsequently James Holmes's second wife); their son Perry became a Baptist minister, preaching his first sermon in Westford in 1869 (see 5/9 entry). Holmes had bought part of his new farm from Dolly Bentley on April 28, 1868 (Otsego County Deeds, lib. 149, p. 514), for $400.

Perry C. Bentley (WN, 1842–1877) was the son of Baptist Deacon Russell Bentley and his wife, Phoebe. The Westford congregation at its March meeting—when it also accepted the resignation of Holmes as clerk—voted to invite young Bentley "to labor for and with this church" as a kind of interim minister (Baptist Book B, 3/6/1869, NYSHA). As Holmes recorded the previous November, the congregation had voted to license Bentley as a preacher at that time. During the school year ending in mid-1870, however, Bentley apparently was attending classes as a theological student at Madison University (FHC.72: 439), and on November 5 of that year he was given a letter of dismission and recommendation from the Westford congregation, addressed to the Baptist church at New Woodstock, near Cazenovia in Madison County. Subsequently he served as minister in the Baptist church at Richmondville, just across the border from Worcester in Schoharie County. He was back in Westford from late 1874 to May 1877, when he received a letter in preparation for moving to Mount Vision in the town of Laurens, but he died on the 18th of that month when he was thrown from his wagon.

Stephen R. Bentley (WN, 1843–1880) was a son of the prominent Westford Baptists Deacon Russell and Phoebe D. Bentley and brother of the minister Perry C. Bentley. He himself joined the church in 1862, as Holmes recorded in his capacity as clerk, but he left it in 1864 and does not appear in the Westford census schedule in 1870 (although he is mentioned in Holmes's 1869 diary and did rejoin the church that year, as well). He was again listed in the 1875 state census as a carpenter in the town. By 1880 he had become simply a farmer. His wife, Daphnea V. Cook, was to be James Holmes's second wife.

"Squire" Nathan Bidlake (WN, 1810–1899) had been married to Emily Kelso (ca. 1824–1865), a cousin of James Holmes. The Bidlakes lived with Emily's parents, David and Mary Kelso (Holmes's uncle and aunt) for most of their married life, so it is likely that Holmes, who also had some land dealings with Bidlake in the late 1860s, knew him relatively well. In his earlier years, Nathan was a farmer; in the 1870s, however, he was a merchant in the village. He was town supervisor in 1846 and served as a justice of the peace (hence the honorific "Squire") from 1850 until at least 1878.

Charles Henry Bissell (1828–1914), a native of the town of Hartwick, had moved to Westford with his wife, Caroline, in 1851. At first they farmed, but he had become an egg-dealer by 1870 and was at that time living in the village. Active in town affairs, he served as supervisor in 1871 and 1874.

H. A. Bolles, M.D. (b. ca. 1826), of Cortland, was in the Westford area in 1869 giving a lecture series on anatomy, physiology, and hygiene. A report from Worcester in the *Schenevus Valley News* on October 26 said that the series was giving "the most complete satisfaction." When Bolles later closed the series at Schenevus itself on November 17, the paper called the experience "a rare treat" and it printed a letter of thanks from some thirty-two prominent citizens.

The December 9th issue of the *Cooperstown Freemen's Journal* included the letter of appreciation along with this puff from G. L. Gates, M.D., who served as assistant to Bolles:

> "To know yourself is the best insurance you can have,
> Against ills, that easier prevented are than cured."
>
> H. A. Bolles, M.D.,
>
> In a series of nine lectures upon Anatomy, Physiology and
> Hygiene, will simplify and explain how to support, nourish and
> preserve in good tone, this fickle frame, so that to live shall be
> a delight, to breath[e] a pleasure, to think a luxury, and to sleep rest.
> A fire or a flood have a beginning, and so has every complaint.
> It is easier to prevent a fire than to put one out. Nine-tenths of our
> diseases result from ignorance of the laws of our being. Come then
> and learn how to shun what you dread to suffer.
> These lectures will be fully illustrated with a costly selection
> of models, skeletons, a Manikin and a dried man,
> besides 1500 feet of oil paintings.

The first lecture in the series was free.

Daniel Brownell (b. ca. 1825), a native of Greene County, had moved to Westford sometime in the late 1850s and was presently farming one hundred acres south of the village with his wife, Loesa, a native of England.

"Dr. Campbell of N.Y.": According to the annual listings of churches in the *Congregational Quarterly*, 11 (1869) and 12 (1870), there was no Dr. Campbell serving a Congregational church in New York City. Nor has any obvious candidate from another denomination emerged from that city. Perhaps Holmes wrote N.Y. as a slip for N.J. and was intending a reference to the widely known William Henry Campbell, D.D., LL.D. (1808–1890), formerly minister at the Third Reformed church in Albany, then principal of the Albany Academy, and in 1869 president of Rutgers University. See *Dictionary of American Biography*.

"F. Campbell": Fenimore Campbell (b. ca. 1840) and his father, Francis (b. ca. 1798), both were farming in Westford in 1869; to whom Holmes refers on January 26 is unclear.

Alpheus B. Chapman was Holmes's brother-in-law from Illinois; see "The Holmes Family," above.

"Chase & Mills": Samuel T. Chase (1817–1876), a native of the town of Maryland who had been engaged in business in Albany for some years and had served for a time as a lumber inspector in Otsego County, was in 1869 running a hardware store in Schenevus with his son-in-law John Mills (b. 1828). Mills, a former wagonmaker, had moved to Schenevus in the 1840s; his relationship with Chase dated from at least 1860, when the census shows him living in Chase's house. But it was the marriage of Mills to Chase's daughter Jennie that led to the business partnership. Following Chase's death, the firm became known as Mills & Gleason. It was taken over in 1889 by John and Jennie's son R. C. Mills. No record of the burial of any of the Mills family has been found in the Maryland vicinity. BR.93: 842–43.

Dwight Chester (WN, b. ca. 1835) was the son of cabinetmaker Alden Chester and his first wife, Mary Chappel, who died in 1837. He was working with his father as a cabinetmaker in Westford in 1850, and after Alden's death was a sash-and-blind maker in the Westford factory owned by Milton Wright's son Henry and James Holmes's brother John. After serving as town clerk in 1859 and supervisor in 1861, he left the area and for many years lived in Boston, where he was working in 1869 as a manager of the Aetna Life Insurance Company. When his half-brother Horace, who had been a banker in partnership with J. T. Thompson in Schenevus, expanded his Bank of Worcester in 1884, Dwight was associated with the new endeavor. This bank later moved to Schenevus following the failure there of the successor of Thompson's bank. BR.93: 335–37.

Harriet Chase Coats (WN, 1820–1906), widow of the farmer Harman (or Hiram) Coats (1812–1867) and later the second wife of the carpenter William Wales, was "keeping house" in Westford village in 1870. By 1875 she was running a tailoring business there.

Walter C. Coats (b. ca. 1856) was the son of Harriet Coats and in 1870 was a fourteen-year-old farm laborer.

William N. Cobb (b. ca. 1818) was appointed presiding elder of the Otsego District of the Methodist Episcopal church in May 1869.

William E. Cory (ca. 1827–1890) ran a hardware store in Cooperstown.

Arthur Cossart (WN, 1851–1921) was a farm laborer living with his widowed mother, Lucy, in 1870. Two other Cossarts—Arthur's brother Ansel Perry Cossart (WN, 1844–1921) and their kinsman Arthur B. Cossart (WN, b. ca. 1845)—to whom Holmes conceivably might have been referring on October 14 probably had left the area by 1869.

Stephen J. Cottrell (b. ca. 1820), whose joining of the Westford Baptist church (by letter) Holmes had recorded in June 1867, had established a cheese factory in the village in 1868. Later that year, when Holmes resigned as church clerk, Cottrell was elected in his stead. Cottrell in turn resigned from this post in April 1870 and, with a traveling letter from the church, apparently left the town soon after. By the time of the 1870 census, only his wife, Betsey, and mother, Matilda, were listed. In August of that year, Cottrell and his wife were granted letters of dismission from the congregation.

"J. Cross": This is almost surely the Joshua Cross who joined the First Baptist church of Worcester (located in East Worcester) by letter in 1858, left it with a letter on May 4, 1867, joined the Westford Baptist church on June 1, 1867, and was granted a letter of dismission and recommendation on March 23, 1873. Although Cross was not listed in the 1870 Westford census, Holmes's diary indicates that Cross was working there—or nearby—in 1869 as a shingle-cutter.

"D. Cumming/D. Cummings": Daniel A. Cummings (1798–1883) was listed in the 1870 census as a retired wagonmaker. He and his wife, Esther Thomas Cummings, were living then in the village, but for the first few years after they arrived in Westford in 1854 they had owned thirty acres of land. Another possiblity here is Albert D. Cummings (WN, 1849–1917), the son of George and Sally Cummings, themselves Westford natives. In 1870, he was a twenty-one-year-old farmer.

A. E. Daniels (b. ca. 1803) was licensed to preach in the town of Morris in 1833 and in 1842 was the organizing minister of the Schenevus Methodist Episcopal church. He was serving at the latter place in 1860 but did not have a local assignment in 1869.

Charles Davis (b. ca. 1808), a native of Rensselaer County who moved to Westford in 1853 with his wife, Grisella, farmed there until at least 1880. In 1870, the Davis farm lay north of the village. One or the other of the Davises may have been related to the Shellands; at this time John Shelland was living next to them, so the visit described by Holmes on March 3, when he took his in-laws, David and Marinda Shelland, to Davis's farm, may have been double.

"Bishop Duan, D. D.": William Croswell Doane (1832–1913) had been elected to the newly created post of bishop of the Albany diocese of the Protestant Episcopal church in December 1868 and was consecrated on February 2, 1869. Just at the beginning of his long career as an influential churchman, Doane was visiting for the first time the Otsego County churches that were under his supervision. See *Dictionary of American Biography* and *Encyclopedia of Living Biography of New York*, vol. 2 (New York: Atlantic Publishing and Engraving Co., 1882), 351–52.

Fernando P. Draper (WN, b. ca. 1820), brother of Susan D. Chester and Milton Draper, ran a hardware store in Maryland village prior to 1868, when another brother, Lysander, took it over. Fernando had worked as a cabinetmaker in Westford in 1850, was listed as a painter there in 1855, and as a mechanic in 1860. In the mid-1860s he had been running a village hotel that catered to the students at Milton Draper's Institute. After selling his hardware business, he returned to the village, where he became a produce-dealer.

Hazer E. Draper (WN, b. ca. 1820) farmed with his father, Dexter (1786–1860), from the 1840s on and later worked his own thirty acres near the village. He last appeared in the census in 1880 and may then have moved away.

Milton Draper (WN, 1829–1877), a brother of Susan Chester and F. P. Draper, was farming in Westford in the mid-1860s. In 1859, he and his wife, Jerusha, apparently had founded what became the Westford Literary Institute, and she continued to teach in Westford until at least 1865. By 1870 they seem to have left the town. (The 1865 state census indicates that the "Draper Institute" had been founded in 1862, but the *Institute Herald* of 3/17/1866 gives 1859 as the founding date.)

Susan G. Draper (WN, 1811–1892) was the second wife of cabinetmaker Alden Chester. Following her husband's death in 1857, Susan

Chester continued to live in their house on the northwest side of the main street in the village. Here was located a large barn that Holmes apparently was renting in 1869, perhaps as a temporary storage place before his own new outbuildings were completed.

"Farry's": Elijah E. Ferry (1812–1877) was a Schenevus lawyer, landowner, and petty industrialist, who in the 1870 federal census reported real estate worth $20,000 plus $8,000 in personal property. An Otego native, Ferry was a harnessmaker when he first came to Schenevus in 1836. Studying law in his spare time, he rose to be a county judge and a member of the legislature and was long associated with the growth of Schenevus. After the coming of the railroad to that village, Ferry developed land he owned between the old main street and the depot. His son Julian, aged thirty-one in 1870, was then superintendent of Ferry's saw, grist, and plaster mills.

"Elder Fox": Hubbard Fox (b. ca. 1818), a Methodist Episcopal minister who was to leave Westford for Schenevus in 1870, was well known locally as a temperance lecturer.

"Free": Evidently a laborer who regularly worked for William T. Marks but who spent May 8 readying Holmes's field for the potatoes he was to plant the next week. A guess would be F[reeman?]. M. Guernsey, a nineteen-year-old farm laborer listed in the 1870 census but otherwise untraceable. The Decatur carpenter James Freeman, twenty-six years old at the time of the 1870 census, is a less likely possibility.

John Goodenough (b. ca. 1830) had moved to Westford by 1867, when he was assessed taxes on ten acres of land there. The 1870 federal census lists him as a farmer. A widower with six children in his house at that time, by 1875 Goodenough had left the area. In 1869, Goodenough did some heavy labor for Holmes.

George E. Gove (1828–1907), in 1869 a Westford farmer, had moved to the town from the village of Maryland in 1862. Holmes had recorded his joining of the Baptist church (by letter) in May of that year.

Daniel M. Greene (1804–1872), who had been working as a carpenter in Cherry Valley in 1850, later moved to Middlefield. In 1869 he worked one day helping Holmes on the Ashley job (9/8 entry); perhaps he owed Holmes some labor in return for work Holmes had done for him.

Arthur Irving Griggs (WN, 1835–1915), the brother of Melville Griggs, took over his father's store in Westford following his graduation from the Richmondville Seminary in Schoharie County. In 1870 he was a "retired merchant" of thirty-five in Westford and owner of real estate worth $8,000 and personal property worth $10,000. Later he seems to have moved to New York City (the *Schenevus Free Press and Davenport Standard* [6/6/1900] places his wife there).

Melville Griggs (WN, 1844–1922), Arthur's younger brother, was town clerk in Westford in 1869 and supervisor the following year. His general store stood on the southwest corner of the intersection between Elk Creek road and the Middlefield or Clarksville road, directly in front of James Holmes's old house. A "retired drygoods merchant" of twenty-five in Westford in 1870, Griggs later moved to Worcester, where he served as supervisor in 1891 and where he died and was buried in 1922.

Harrison P. Groff (b. ca. 1838), son of the Westford native Archibald Groff, was not listed as a resident of the town in the 1870 census but had returned by 1875, when the state census listed him as a farmer.

Nathaniel Groff (WN, 1803–1874) owned 185 acres of land in 1858; this had dwindled to 60 acres by 1863. His nephew and namesake, Nathaniel (WN, 1841–1896), was by 1870 owner of 135 acres. Both were farmers; to which of them Holmes refers on January 6 is unclear.

Samuel H. Gurney (b. ca. 1828), listed as a "huckster" in the 1870 census for Maryland township, had run a general store in Schenevus in partnership with George L. Cass until March 1869. Since that time, when Cass left Schenevus for Gorry, Pennsylvania, Gurney had been running it alone. Sanders J. Gurney (b. ca. 1840) was listed as a carpenter in the 1870 census. It seems likely that Holmes purchased something from Gurney's store on October 29, but it is possible that he settled some account with his fellow craftsman.

Robert S. Hall (b. 1837), a native of Roxburghshire, Scotland, immigrated to America with his parents in 1854. He worked on his parents' farm until 1859, when he began learning to be a miller. In the 1860s he purchased Elk Mills from Myron J. Hubbard. At his mill adjacent to

Holmes's new house, Hall produced lumber as well as grain. George Guy, one of the mill's previous owners, reported in the 1860 census that he had turned out 120,000 board feet in 1859, probably custom work as well as production for the local market. It seems clear that Hall was doing custom sawing of timber or rough lumber for Holmes in 1869. The wood was almost certainly hemlock (memoranda 1 and 6) purchased locally for Holmes's own house. Although Hall was still listed in the 1900 federal census as running the Elk Mills, and the death of his wife, Christie Bell Hall, also a native of Scotland, in Westford was reported later that year in the Schenevus paper, neither of them seems to have been buried in any Westford cemetery. Hall served Westford as town clerk in 1877 and as supervisor in 1878, 1879, 1890, and 1891.

David S. Herdman (WN, b. ca. 1840) was the half-brother of Holmes's helper Everett Herdman. He was still living with his father and stepmother in 1870, and OTCO.72 suggests that he was still farming with them. He does not show up in the 1880 census.

Everett D. Herdman (WN, b. ca. 1852) was a young worker hired by Holmes as his helper in April 1869. His father, John Herdman, had been listed as engaged in manufacturing and trades in the 1840 census in Westford and in 1850 was a harnessmaker. In 1855 and later years, however, he declared himself a farmer. See chapter 1.

"Rev. Hobs/Dr. Hobs of NJ": This visiting minister, whom Holmes heard on May 23 and on July 25, has not been identified further.

Betsey Hubbard (b. ca. 1828) was the second wife of Myron J. Hubbard. She apparently left Westford sometime after his death in 1888, for she does not show up in any local cemetery records.

Charles A. Hubbard (b. ca. 1855) was the younger son of Myron J. Hubbard by his first wife, Lavancha, who died in 1863. By 1875 young Hubbard had left Westford.

John Hubbard (b. ca. 1834) operated a Schenevus brickyard that burned in July 1869. In the summary of expenses on his own house, which Holmes entered at the end of the diary, he listed the cost of four hundred brick as $6. A "Cash Account" entry on March 1 ("John Hubbards by S. Manzer, $6.00") suggests that Hubbard had supplied the bricks in January.

Myron J. Hubbard (WN, 1828–1888), colonel of the thirty-ninth regiment of the New York State National Guard, spent much of his life farming, with intervals during which he was a schoolteacher, a merchant, and the owner of the mill in Westford purchased from him in the 1860s by Robert S. Hall. Hubbard served Westford as supervisor in 1854 and as town clerk in 1860 (the year before Holmes), was elected county coroner in 1862 and justice of the peace in 1862 and 1866, and won a seat in the New York assembly in 1867 as a Democrat. In 1869, Holmes was building Hubbard a new house in Westford. See S. R. Harlow and S. C. Hutchins, *Life Sketches of the State Officers, Senators, and Members of the Assembly of the State of New York, in 1868* (Albany: Weed, Parsons, & Co., 1868), pp. 251–52.

John Hull (WN, 1825–1887) and his brother Irving (WN, b. ca. 1839) were farming jointly near the Holmeses on the Worcester road. When Mary Holmes picked hops for them late in August, she was taking part in an annual occasion in Otsego County, one of the main hop-raising areas of the nation in this period. The *Schenevus Valley News* reported on August 24 that the crop was "very heavy" this year and that picking would begin that week. On August 31, it noted that for the very first time the new railroad was bringing large numbers of city-dwellers to Otsego County to help in the harvest.

David Kelso (1802–1871), the brother of Mary Kelso Holmes and thus James's uncle, listed himself as a retired blacksmith in the 1870 census. But in those for 1850 and 1855 he is called a farmer and in 1860 and 1865 no occupation is given for him. He served Westford as supervisor in 1848, 1853, and 1855–1856 and as town clerk in 1843.

James Kelso (WN, b. ca. 1796), also a Westford blacksmith for many years, was the brother of Holmes's mother, Mary. Kelso is listed in the 1865 state census, and his house appears in the 1868 county atlas, but he is not listed in the 1870 Westford census and apparently was not buried there. An important source of financial support for his nephew (and namesake?) James, he was to lend him $200 in December 1869.

Mary Prevost Kelso (ca. 1804–1875), who was married to David Kelso, was one of six sisters who came from a rather unusual social background. Among her relatives by blood and marriage were the Franco-British officers Gen. Augustine Prevost and Lt. Col. James Marcus Prevost; the lawyer, public official, and diplomat John Bartow Prevost; and Aaron Burr and his daughter Theodosia, whose mother was Mary Prevost Kelso's grandmother. Mary's father, Augustine James Frederick Prevost, and his second wife moved the family to Westford around 1820; four of his daughters married there and remained when the rest of the family moved to Illinois around 1840.

"Kise's Hotel": Daniel L. Keyes was listed in OTCO.72 as proprietor of the St. James Hotel on Main Street in Cooperstown; Webster C. and Squire W. Keyes were listed in the same directory as proprietors of the Central Hotel, also on Main Street.

Freeman P. Lang (b. ca. 1820) had been a schoolteacher in Deerfield township, New Hampshire, in 1850. In the summer of 1856, he came to Westford to serve as minister for the Baptist church. He resigned from the church in April 1859 in protest over the inability of its members to pay his salary, but he had forgiven the oversight sufficiently by 1862 to revisit the church then (as Holmes noted in the clerk's records) as well as in 1869.

Rev. Delmer R. Lowell (b. ca. 1844), a native of Schoharie County, had lost his right arm at the battle of Sailor's Creek in the Civil War. Lowell had resided in Westford before entering the army; in 1865 his parents still were living there. Hence his visit to the Methodist Episcopal church at this time was something of a homecoming.

David McClintock (1827–1875), a native of Delaware County, moved to Westford in 1853. There he farmed north of the village until his death, which occurred when he "fell and broke his neck in the woods" (so reports the state census) in January 1875. His wife, Abigail Snyder (WN, 1830–1915), was a member of a family that produced several Westford carpenters; she took as her second husband the carpenter Oliver Thurber.

Leander Manzer (WN, b. ca. 1830) was the younger brother of Sherman Manzer. He and his wife, Maria, had been farming in the area since 1850, but by 1870 they had left Westford.

Sanford E. Manzer (WN, b. ca. 1826) farmed in Westford from the late 1840s until the early 1860s, when he became a grocer, housing his business in the building previously used by George Skinner, across the street from Melville Griggs's store. Serving Westford as town clerk in 1862, 1867, 1871, and 1874, he left the area sometime thereafter.

Sherman Manzer (WN, 1824–1882) called himself a farmer in the 1870 census. Holmes knew Manzer well: the 1865 census shows Manzer and his wife, Mary, living with Holmes's family in the latter's house.

Leroy Marks (b. ca. 1838), like William T. Marks a native of Cherry Valley (they were probably cousins), was a farmer in the town of Roseboom in 1872.

William T. Marks (b. 1827), who hailed from the town of Cherry Valley to the north of Westford, had married Holmes's sister Mary Lucy in 1849; see "The Holmes Family" above. Marks was a farmer (with fifty acres) in 1870, and in OTCO.72 was listed as running a cider mill as well (when Holmes noted on October 12 that he "worked for W. T. Marks on water wheel," he in fact may have been fixing or helping build part of the cider mill). Following the death of his wife in 1885, Marks seems to have left the town.

John Mason (1795–1869), born in Connecticut, had come to Westford with his parents in 1798. He and his wife, Elizabeth (1792–1865), farmed and raised thirteen children there. In 1855, James and Mary Holmes had been near neighbors of the Masons north of the village.

N. S. Moore, a Congregational minister, was listed in OTCO.72 as at that time serving the Westford congregation.

Michael S. Murphy (1825–1900), a native of Columbia County whose parents came to Westford around 1850, first appears in the 1858 assessment book and in the census in 1860. In 1870, he was farming northeast of the village.

Robert Pank (1810–1887), a Westford farmer, is probably the man for whom Holmes filed a saw in January, although one of the "Pank boys"—William (b. ca. 1839), Matthew (1841–1896), or Joseph (1844–1872), three of Robert's sons then living in Westford—may be the person indicated.

William Pank (b. ca. 1839) was a sometime carpenter in Westford; in 1870, the census shows him living with his parents, Robert and Mary Ann, but lists no occupation for him. OTCO.72 lists him as a carpenter and farmer. The 1870 census mortality schedule reports the cause of death for his wife, Mary Jane Pank (b. 1842; see 10/7 entry), as consumption, but childbirth may have contributed to it.

"Parkers": Catherine (1824–1882) and *George W.* (1815–1876) *Parker* of Decatur may be the persons Holmes referred to on March 27, in which case the former probably was Anna Howe Holmes's daughter by Artemas Howe. George W. Parker was a farmer, dairyman, and hop-raiser in Decatur and a member of a large family centered in that town.

George Parsons (1811–1900) was the minister at the Worcester Methodist Episcopal church in 1868–1869. He had started his ministry in 1838, was in nearby Decatur in 1850, and had been in Schenevus in 1863–1864. He was reassigned to Fly Creek in 1870 and was listed as still there in OTCO.72. (See a report of his death, at Ft. Plain, on May 7, 1900, in the *Schenevus Free Press and Davenport Standard*, 5/16/1900. The substitution of *J* for *G*, as in the diary on July 18, was common at the time.)

Thomas Patrick (WN, 1836–1916), a farmer and dairyman, was listed in OTCO.72 as working 150 acres south of Westford village. His wife was Sarah A. Chase.

"H. Peobles and Jno.": Hugh Peebles (1802–1884) was a farmer worth $32,000 in 1870; John E. Peebles (b. ca. 1833), apparently his son, was a carpenter in Schenevus in the same year.

William H. Platner (WN, 1835–1892) was a Westford wagonmaker and carriagemaker and an undertaker. He was elected township supervisor in 1868 and 1869. Platner's wife, Amanda, was the sister of merchants Waldo and Ken Tyler.

Ingraham Powers (b. 1811), a Worcester native, attended Madison (Colgate) University from 1835 to 1838, was ordained as a Baptist minister in the latter year, and spent much of his active life in Otsego County. He was pastor of the Second Baptist church in Worcester from 1863 to 1871 but then moved to Seward (Schoharie County) and Decatur. See FHC.72: 243.

Leonard F. Preston (WN, b. ca. 1819) appears as a farmer in each census through 1865, but in 1870 he is listed as a bootmaker and shoemaker. He and his wife, Mary Amanda, apparently had left the area by 1880.

John L. Ray (b. ca. 1840), in 1869 pastor of the Baptist church at nearby Westville (which both Holmes, in the 3/21 and 9/5 entries, and Milton Wright, in the clerk's records on March 1, 1851, call Waterville), had entered the class of 1866 at Madison (Colgate) University but did not graduate (FHC.72: 477). During 1869, he officiated at worship services in the Westford Baptist church on three occasions. He left Westville in April of 1870, shortly before the census was taken, and has proved untraceable thereafter.

"Fanny Roberts": Fanny North (1840–1920), originally from Middlefield, and Fannie Platner (b. ca. 1833), of Westford, married the brothers George Nelson Roberts and William Henry Roberts, but the reference here seems not to be to either of them. Apparently Fanny Roberts of Illinois was the daughter of Anna Howe Holmes, the second wife of Calvin. In May, Leander Roberts, who must have been her husband, came to Westford "after [his] mother." Presumably it was Holmes's letter, mentioned on March 24, that brought Roberts's husband to the village two months later.

George Nelson Roberts (WN, 1834–1915) was for a long period a teacher in Westford schools, as well as superintendent of the Methodist Sunday School for nearly forty years, and was supervisor in 1888, 1889, and 1892. He also farmed; he owned some 212 acres from the mid-1860s at least up to 1875 and probably well beyond.

Leander Roberts apparently was Anna Howe Holmes's son-in-law. He visited Westford in May 1869 from Illinois, perhaps to take the recently widowed Anna back home with him.

"Rose": Edward A. Rose (b. ca. 1839), a farmer who lived about a mile from Holmes's new house on the Worcester road and who probably was known to Holmes through the Baptist church, seems to have been the person referred to by Holmes on May 20. He and his wife, Martha, and his father, Harman, had moved to Westford in 1867 or 1868, probably from Rensselaerville in Albany County. But the farmer John J. Rose of Maryland village (probably no kin; b. ca. 1822), who ran a lumber and materials business there that had been established by his father, Nathan, many years earlier, is also a possibility (see his records books in NYSHA). The case for John J. Rose may be especially strong with regard to the lime Holmes paid "Rose" for on July 16 (note,

though, that the lime he himself purchased on May 15 in Schenevus he almost certainly bought from Chase & Mills, not from John J. Rose).

Harman Rose (ca. 1792–1869) was the father of Edward A. Rose. Although neither Edward nor his father was a member of the Westford Baptist church, Baptist minister Ingraham Powers came from Worcester for Harman's funeral. Edward's wife Martha A. Rose, however, was a member of the church; she had joined in March 1868 by a letter from the Rensselaerville congregation.

H. O. Rowland (b. ca. 1848), a native of Wales, was in Westford during the summer of 1869 because he had been induced to preach for the Westford Baptists during his vacation from Madison (Colgate) University. In 1870, Rowland was listed in the census for Middlefield, where he was to serve the Baptists until replaced in 1871. By 1873, he was located in Whitestown, Oneida County, where his wife, Louisa Howland, a Westford native, died in that year.

Denton Salisbury (WN, 1823–1893) had begun his career as a blacksmith and then worked as a wagonmaker before becoming a stonemason in the late 1850s. He was still practicing that craft in 1870, three years after his wife, Amanda, died; he and his children were then next-door neighbors of the Holmeses. In 1869 he apparently was the collector of local school monies.

"Seeley's": David W. Seeley (1821–1903), a native of Schoharie County, apparently moved with his father, Oran, to the Albany area in the 1850s. Albany directories indicate that the elder Seeley was a patent-rights owner or dealer and farmer; his son was variously partner in a stove store, a patent-rights broker, supervisor of the Fifteenth Ward (in 1870), a gas-and-steam fitter, a general broker, and then patentee for a "perpetual hay and cotton press" or baler, which he continued to sell from his brokerage on State Street until at least 1883. By 1900, he was a farmer on Western Avenue. When Holmes visited the Seeleys in 1869, they were living on Madison Avenue at the corner of Ontario; David was then a patent-rights broker. How Holmes became acquainted with Seeley is not known; in 1850, Seeley was listed in the federal census as a resident of the town of Carlisle, Schoharie County, fairly close to Westford. Some kinship link seems reasonable, although this certainly was not very close: following the death of his first wife,

Maria, Seeley married (on September 23, 1891) Holmes's sister Clarissa. That Holmes's relation to Seeley was more than casual in 1869 is suggested by the fact that when the Seeleys later visited Westford they spent a good deal of time with the Holmeses.

"Father and Mother Sheldon": David Shelland (ca. 1795–1869) and *Marinda Shelland* (b. ca. 1800) were the parents of Mary Shelland Holmes. David was the brother of Elisha Shelland and (probably) John Shelland. David Shelland died in late March 1869. See "Mary Holmes" in "The Holmes Family" above.

Elisha Shelland (1790–1878) was the uncle of Mary Shelland Holmes. For much of his life a farmer in the town of Decatur, just east of Westford, he had lost his wife in September 1868 and in 1869 he apparently was living with (or at least boarding with) James and Mary Holmes. His presence with them was brief; the death of Mary's father brought her mother, Marinda, into the household by the end of 1869.

John Shelland (b. ca. 1801) probably was the brother of Elisha Shelland and hence another of Mary Holmes's uncles. In 1870, John and his wife, Anna, were living on a farm north of the village.

George Skinner (1792–1875), a native of Woodstock, Connecticut, had moved to Westford in 1816. There he served as a merchant; as postmaster; as town clerk from 1824 through 1834 and again in 1841, 1842, and 1844; as supervisor in 1843 and 1845; and, for fifty years, as librarian of the Union Library. He succumbed to heart disease at the age of eighty-three in April of 1875.

"Skinner": James B. Skinner (1819–1905), husband of Holmes's sister Fanny; see "The Holmes Family" above. Three of Skinner's brothers, all farmers, are mentioned in the diary: John R. (1811–1877), William (b. ca. 1827), and Zebadiah (1822–1914). The family occupied the general region of Skinnerville north of the village and just west of the Calvin Holmes farm.

"H. J. Smith": Rev. H. A. (not H. J.) *Smith* was the Financial and Corresponding Secretary of the Educational Society, which supported students at Madison (Colgate) University in Hamilton, New York, a Baptist college founded in 1819. He had been canvassing in Schenevus the week previous to Holmes's mention of him on January 24. See *Minutes of the 74th Anniversary of the Otsego Baptist Association* (Cooperstown:

Freeman's Journal Office, 1869); *Schenevus Valley News*, 1/19/1869, p. 2; FHC.72: 408.

Andrew (or *Andrews/Andrus*) *Snyder* (1796–1889), a native of Rensselaer County, had come to Westford just after 1800 with his widowed mother, Elizabeth, and her many other children. Long a carpenter, he was related to James Holmes via marriage: the sister of his first wife, Louisa Prevost, was the wife of Holmes's uncle David Kelso.

Erasmus Snyder (WN, 1822–1898) was the nephew of carpenter Andrew Snyder, from whom he probably learned the craft. He was practicing it in the town of Middlefield in 1850 and did not return to Westford until the late 1860s; in the 1870 census population schedule, he appears next to James Holmes's brother John.

John A. Snyder (WN, 1821–1899) was the son of carpenter Andrew Snyder. Long a blacksmith in Westford, he had several links with the Holmes family: in 1850, Anna Howe (soon to become James's stepmother) was living with Snyder and his wife, Mahala Bates; their daughter Susan, in turn, was to marry Revilo Holmes, James's nephew; finally, John A. Snyder's aunt Mary Prevost was married to James's uncle David Kelso.

Lyman Andrew Snyder (WN, 1827–1912), a nephew of carpenter Andrew Snyder, farmed south of the village.

"Rev. Swinerson": Henry Ulyate Swinnerton (1839–1927) had become minister of the Presbyterian church of Cherry Valley in May 1868. He probably conducted the services in the Westford Congregational church on July 25. See his own pamphlet, *An Historical Account of the Presbyterian Church at Cherry Valley, N.Y.* (Cherry Valley: Gazette, 1876).

James R. Thompson (b. ca. 1846) was a cooper briefly resident in Westford village in the 1860s.

John T. Thompson (b. ca. 1819), who served the town of Maryland as both supervisor and clerk on a number of occasions, opened his general store in Schenevus in 1864 but had been listed as a merchant there as early as the 1850 census. In the 1870 census, he was listed as a farmer of fifty, owning real estate worth $55,000 and personal property worth another $15,000. It is likely that Holmes bought flour from Thompson's store on his trip to Schenevus on February 25, although it is conceivable that the Westford cooper James R. Thompson (see 8/25 and 9/1 entries) supplied it.

Jacob Tipple (b. ca. 1822) was a laborer who had recently purchased the house just across the road from Denton Salisbury, Holmes's new next-door neighbor. Holmes evidently was renting Tipple's barn for hay storage during 1869.

William Henry Tipple (1783–1869), long a farmer in Westford but not a resident for many years, died on December 27, 1869, and was buried in Worcester.

Elisha R. Treat (WN, b. 1829) in 1870 was running a butcher shop in the village.

Anson Tucker (b. ca. 1825) was a Pennsylvania native who had lived in Westford in the early 1850s. He then left, only to return again by 1860. In 1870, he and his wife Sophia were living a short distance north of the village; living with them were their daughter Ada and her husband, Asa Butler, then working as a carpenter.

Charles B. Tyler (WN, b. ca. 1848), the brother of Westford merchants John K. and Waldo H. Tyler, was farming north of the village in 1870. Later he and his other brothers, Oscar and Azro, moved to Ohio.

Menzo Tyler (WN, 1835–1914) was the first cousin of Waldo, Ken, Charles, and Oscar Tyler. After the suicide of his father, James Tyler, in 1864, Menzo ran the family farm with his mother, Samantha. In 1870, he was still unmarried and was living with her and her other children south of Westford village.

Oscar H. Tyler (WN, b. ca. 1841), brother of Waldo, Charles, and Ken Tyler, was a farmer—and, on evidence of the diary, a teamster—who may have owned the quarry from which the stone for Holmes's house was taken.

"Tyler Brothers": Waldo H. Tyler (WN, b. 1839) and his brother *John Kendrick Tyler* (WN, b. 1843) were running a general store together in Westford in 1869. Waldo entered that business in 1860 as the partner of J. T. French, with whom he continued until 1862; sometime thereafter he took his brother "Ken" into the business and then, buying him out in April 1869, continued to run the store until the 1890s, perhaps

even later. Ken Tyler began an insurance agency affiliated with the Glens Falls Insurance Company in Westford in July 1869. Both men served Westford modestly in public office, Waldo as supervisor in 1865 and 1867 and Ken as town clerk in 1872 and 1873. But eventually both left the town. Ken moved his insurance agency to Worcester, where he became affiliated with the Aetna Company, and Waldo relocated to Schenevus, where he became a village trustee. The dealings between Holmes and the Tyler Brothers in 1869 were extensive. Most importantly, they paid him a total of $243.50 over the year. This suggests that they were lending him cash for his operations or that they were paying him for some sizable structure he had built them in 1868. BR.93: 260–61.

William Utter (b. ca. 1807), a native of Green County, was a farmer who with his wife lived about a mile east of the village.

"Vanhoser": Henry N. VanDeusen (b. ca. 1840) was the Methodist minister in Schenevus in 1868 and 1869. Early in 1870, he was reportedly reassigned to Laurens, although the census for that year does not list him there.

Dr. Erastus B. Warren (1831–1897) was a homeopathic physician in Clarksville (now Middlefield), the next village northwest of Westford.

"M. Waterman": John M. Waterman (b. ca. 1831) was listed in the 1870 census in nearby Decatur as a carriage-painter. Apparently a resident of Westford in 1867, when "J. Marvin Waterman" appeared in the assessment book, he seems the likeliest of several possiblities as the supplier of paint to Holmes on April 10.

William O. Wilbur (WN, 1822–1892) was a farmer who owned land adjoining the parcel Holmes repurchased from Daniel Aldrich in March.

Judson Winans (b. ca. 1839), Holmes's most skilled assistant in 1869 and one of his clients as well, was the nephew of Milton Wright. He came to Westford as an orphan in 1855 and was to leave the township for Oneonta in 1870. After staying there for some years, he moved to Worcester. See appendix 2.

George Wing (1826–1903), who was farming 126 acres near German Baldwin north of the village in 1870, had been operating a green-tow (i.e., linen) mill in the village proper in 1860, but apparently he had left the area for part of the interim. He served as town supervisor from 1875 to 1878.

Cornelius Winne (1816–1884) and his wife, Margaret, moved to Westford with their six children sometime between 1860 and 1865. They at first lived in the village, but by 1867 they had acquired a farm near that of William and Lucy (Holmes) Marks north of the village.

Chester Wright (ca. 1805–1871) and his wife, Irene (b. ca. 1811), had several ties with James and Mary Holmes. Chester Wright, a farmer, probably was Milton Wright's older brother; certainly in 1850 the two were very near neighbors, along with several other of Milton's known relatives. Living with Chester and Irene Wright at that time was Mary Shelland; following her marriage to James Holmes five years later, the new couple lived with the Wrights. Living there also, both in 1850 and in 1855, was Alpheus B. Chapman, who was to marry Holmes's sister Julia in 1857. A last tie is the fact that in 1870, after the Wrights had retired to Worcester, their domestic servant was Orilla Shelland, Mary Holmes's sister.

Flavel B. Wright (WN, 1826–1907), was half-brother of the carpenter Milton Wright. As the husband of Hannah Skinner (WN, 1833–1913), whose brother James was married to Holmes's sister Fannie, he was kin.

Pomeroy Wright (1787–1876), a Massachusetts native, was one of the two original deacons of the Westford Baptist church at its founding in 1825 and had been a central figure in its life ever since. He was himself a farmer, but numerous of his relatives (including his sons Milton, John, Warren, and Henry) were carpenters. In 1870, although still listed in the census as a farmer, he was living in Westford village with his second wife, Tamison Seward. His daughter Louisa married James Holmes's brother John, also a carpenter, in 1851.

Appendix 2

The Carpenters of Westford: A Biographical Gallery

The biographical entries below repeat schematic information from table 5 (see explanatory key) and table 9 and contain additional data gathered from the sources indicated in "General Note on Sources." Capitalized names in any entry refer the reader to other entries in this appendix. The letters WN identify Westford natives.

1. RICHARDSON BABCOCK (WN: 1797 – 1875)

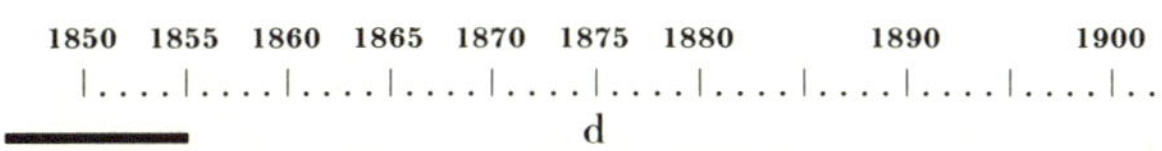

WN: WFD—Worcester—Albany—WFD—Maryland

Already practicing carpentry in Worcester by 1820, by 1830 Babcock had moved to the city of Albany. But his stay there did not last long: Dr. Elhanan Jackson mentions his presence in Westford in the summer of 1834, and in 1840 he was still in the township. He was practicing carpentry there then, as he was in 1850, when, at age fifty-two, he reported owning real estate worth $6,750. Within three or four years, however, he and his wife, Mary, moved to Maryland. There he continued to practice through 1865. Although listed as a farmer in the Maryland census five years later, in OTCO.72 he was said to be a carpenter once more. His daughter Sarah Jane married Westford carpenter WILLIAM H. GROFF.

2. GEORGE BECKER (WN: 1832 – 1893)

1850 1855 1860 1865 1870 1875 1880 1890 1900

d

WN: WFD—?—WFD

George Becker, son of Westford farmers William Becker and Gittie Smith, was himself a farm laborer in the Westville area in the 1850s. Not reported in the Westford census in 1860, by the early 1860s he had bought a lot in the village of Westville, and by 1865 at the latest he had begun to practice carpentry there. The only man still officially active in the craft in all of Westford fifteen years later, he continued to practice as a carpenter up to his death in 1893. His grandfather Rinear Becker (1783–1854) was listed in the 1840 Westford census as a craftsman of some sort; it is possible that young George learned carpentry from him.

3. AARON WEBSTER BENTLEY (b. ca. 1822)

1850 1855 1860 1865 1870 1875 1880 1890 1900

Otsego Co. N: ?—WFD—Oneonta

An Otsego County native who was living as a twenty-six-year-old laborer in Cherry Valley in 1850, Aaron Bentley had moved to Westford and become a carpenter there five years later. He claimed then that he had spent a total of eight years in his new home, so perhaps he had lived there for a few years prior to 1850. In 1860, he was a farmer; within a few years, however, he had left the township. During the 1870s, he was farming in Oneonta township, although as late as 1892, at the age of seventy, he once more claimed to be practicing carpentry.

4. EDGAR BENTLEY (WN: b. ca. 1829)

1850 1855 1860 1865 1870 1875 1880 1890 1900

WN: WFD—?

A descendant of the early Westford settler Rufus Bentley, Edgar had spent all but two years of his life in the township before 1855, when he declared himself a carpenter. Within five years he had left the area, and he seems never to have returned. He left behind his great uncle WILLIAM BENTLEY, long a Westford carpenter, and his uncle

STEPHEN BENTLEY, who practiced briefly before his untimely death.

5. STEPHEN R. BENTLEY (WN: 1843 – 1880)

1850 1855 1860 1865 1870 1875 1880 1890 1900

d

WN: WFD—?—WFD

Son of prominent Westford Baptists Russell and Phoebe ("Dolly") Bentley, Stephen Bentley was the nephew of carpenter WILLIAM BENTLEY and the uncle of carpenter EDGAR BENTLEY. He was working as a laborer in Westford in 1860. Clerk James Holmes recorded his joining the church in May 1862 and also his departure in February 1864. He was gone from the township in 1865 and 1870, although the Holmes diary makes one reference to him in 1869 (and in that year he rejoined the church as well). By 1875 he was back and was practicing as a carpenter there. Five years later he had become a farmer, but in the summer of 1880, at the age of thirty-seven, he died. His widow, Daphnea V. Cook, was to marry JAMES HOLMES in 1881.

6. WILLIAM BENTLEY (1791 – 1862)

1850 1855 1860 1865 1870 1875 1880 1890 1900

d

Columbia Co./Canaan twp. N: Columbia Co.—WFD

William Bentley, a native of Columbia County, New York, migrated between 1810 and 1814 (or perhaps earlier) to the Westford area, where his brother Rufus had relocated in 1798. They were joined there over the following years by a number of other kin, including two more brothers and their parents, Barbara ("Barbery") Vaughan and Thomas Bentley, Sr. In 1814, his surviving account book indicates, William already was practicing carpentry in Westford. He remained active through the mid-1840s as a furniture-maker, repairman, house-builder, and general-purpose woodworker. During the 1850s, however, he switched to agriculture, only to call himself a carpenter once more in his last census entry in 1860. Long resident with his wife, Brachie Dumont, in the area north of Westford village known as Bentley Hollow, William was probably the most important carpenter in the township before 1850.

7. PETER BROWN (1788 – 1874)

1850 1855 1860 1865 1870 1875 1880 1890 1900

|....|....|....|....|....|....|....|....|....|....|..

d

Scotland N: Scotland—Philadelphia—Cherry Valley—WFD

Leaving his native Scotland with his family, Peter Brown had arrived in Philadelphia in the mid-1820s and was naturalized there in February 1828. Appparently he moved soon thereafter to Cherry Valley, an area north of Westford where a number of British immigrants located; in 1830, he was recorded in the Cherry Valley census. The 1840 census did not report him there (or apparently anywhere else in New York State; perhaps he was missed), but in 1850 he was counted again—for the first time as a carpenter. By 1855, in his late sixties by this point, he had moved to Westford and was practicing there. He called himself a farmer in 1860 and 1865, although he owned only a few acres of land; in 1870, four years before his death at age eighty-six, he once more was a carpenter.

8. ASA BUTLER (1848 – 1875)

1850 1855 1860 1865 1870 1875 1880 1890 1900

|....|....|....|....|....|....|....|....|....|....|..

d

Worcester? N: Worcester—WFD—Worcester

In 1850, when young Asa was two, his father and mother, Rosander and Harriet Butler, were farming in Worcester township, as were his grandparents Asa and Betsey Butler. The elder Butlers had lived in Worcester for a long time, but they must have farmed close to the Westford border, for they also had been associated with the Baptist church in Westford village for decades. Asa's grandparents joined the congregation in 1828 by a letter of transfer from Worcester; and their son did so by baptism in 1834. Asa's grandfather, in fact, served Westford's Baptists as an elder from 1828 until the 1850s. When young Asa showed up in the Westford census in 1870, then, as the twenty-two-year-old husband of Addie Tucker (with whose family they were living), he had simply moved across the border into a township where his family—and, presumably, he himself—was well known. The census rather pointedly reported in that year that Asa was not quite a full-fledged carpenter: "Carpenter—works at," says the occupational listing for him. For the next listing, he was once more living in Worcester and now was called a carpenter, pure and simple. But in October he became gravely ill with typhoid fever and, at the age of twenty-seven, died.

9. EDWARD CHASE (WN: b. ca. 1835)

1850 1855 1860 1865 1870 1875 1880 1890 1900

WN: WFD—?

Both of Edward Chase's parents, Pliney and Marilla, had come to Westford as young children—he in 1801 as a five-year-old from Vermont, she in 1801 as a three-year-old from Columbia County, New York. They farmed there through their long lives, raising a large family. Edward, one of the younger children, was fifteen in 1850, living as a laborer with farmer Lester Babcock. Five years later, he was back with his parents but was a carpenter. With him was a five-month-old baby, Austin; Edward's wife apparently had died recently, although no record of her death or burial has been found. By 1860, both father and child were gone from Westford, and neither seems ever to have returned.

10. ALDEN CHESTER, SR. (1802 – 1857)

1850 1855 1860 1865 1870 1875 1880 1890 1900

d

New London Co. (Groton twp.), CT N: New London—WFD

Coming to Westford with his parents as a boy sometime prior to 1820, Alden Chester ran a wagon shop in Worcester for a couple years in the early 1830s and then went back to Westford. He was still in some trade in 1840; by 1850, he had become a cabinetmaker. This was still his occupation in 1855, although an 1856 deed for property in Westford village refers to his "sash and blind factory" there; a biographical sketch of one of his sons published in 1891 says, "At first [Alden Chester, Sr.] was a cabinet-maker, he afterward carried on the business of manufacturing sash, blinds, and doors." His second wife, Susan Draper, whom he married in 1838, was a member of a large and prominent Westford family that produced, among others, the sometime cabinetmaker F. P. DRAPER (her brother) and Andrew Sloan Draper, later president of the University of Illinois and education commissioner for New York State. Susan and Alden's children also achieved considerable prominence: Alden, Jr., as a lawyer and state supreme court justice, HORACE as a banker in Worcester and Schenevus, and DWIGHT (actually Alden's son by a previous marriage) as an executive for Aetna Insurance Company in Boston.

11. DWIGHT CHESTER (WN: b. ca. 1835)

WN: WFD—New York City—Boston

Son of ALDEN CHESTER, SR., fifteen-year-old Dwight was, like him, listed as a cabinetmaker in the 1850 census for Westford. Apparently he left the township for some part of the following decade, although in the later 1850s he was back. Now listed as a sash-and-blind maker, he must have been running the factory that HENRY WRIGHT and JOHN HOLMES had owned in 1855 and that his father apparently purchased from them shortly before his death in 1857. Dwight reported that the business was worth $1,800 in 1860; it used 20,000 feet of lumber valued at $300, was water-powered, and employed four hands (Dwight and his half-brother HORACE, plus HENRY WRIGHT and WARREN WRIGHT must have been the men in question). It turned out 500 sets of blinds (i.e., shutters), 100 doors, 3,000 lights of sash (worth, together, $999), and $600 of "other work." But Chester was not long for the township. Chosen town clerk in Westford in 1859 and elected supervisor in 1861, he had left the vicinity by 1865, going first to New York City and then to Boston, where he became an executive with the Aetna Insurance Company.

12. HORACE CHESTER (WN: b. 1842)

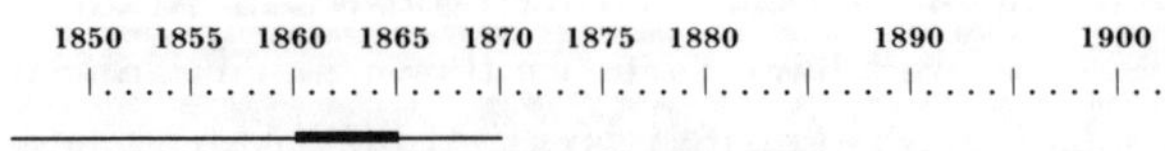

WN: WFD—Maryland—Worcester—Maryland

Only fourteen when his father ALDEN CHESTER, SR., died in 1857, Horace (like his half-brother DWIGHT, seven years his senior) still must have learned some of the craft at home—from Dwight if not from Alden. In 1860, Horace was a sash-and-blind maker at work in the factory once run by his father and now managed by Dwight. By 1865, he was teaching school in Schenevus (the census in 1865 listed him with his mother, Susan Draper Chester, in Westford, but apparently he had moved to the other village by then: BR.93: 335 says he did so in 1864). After teaching school in Schenevus, he became a telegraph operator and the agent there for the Albany & Susquehanna Railroad. He later entered the field of banking there, then in Worcester, then once again in Schenevus. By the 1890s he was a prominent local businessman, and his Chester Banking and Exchange Office was the major financial institution in Schenevus then.

13. ADDELMAN/ADDELMORN CLEAVELAND (b. ca. 1829)

1850 1855 1860 1865 1870 1875 1880 1890 1900

Worcester(?)N: Worcester—WFD—Worcester

A farmer in Worcester in 1850, living with his widowed mother, Hannah, Cleaveland left the region soon afterwards. Perhaps he went to Albany County, where his wife, Adeline, and at least one of their children were born. In any event, he shows up in Westford in 1870 as a farmer, then in OTCO.72 as a carpenter and farmer. Moving back to Worcester by 1875, he called himself a carpenter there in that year and in 1892. In 1878, he filed a mechanic's lien against a resident of Decatur for recovering the cost of his labor in erecting a hop house there. Five years later, still in Worcester, he filed a lien against a resident of that village for work done on a dwelling there.

14. WILLIAM DARLING (ca. 1813 – 1886)

1850 1855 1860 1865 1870 1875 1880 1890 1900

d

Columbia Co. N: Col. Co.—WFD—Decatur

A native of Columbia County, William Darling first shows up, as a farmer, in Westford in 1840. In 1850 he became a carpenter, but by 1855 was gone. He showed up in Decatur in 1860 as a mechanic (a common term for any manual craftworker), then as a carpenter there in 1865 and 1870—only to become a mechanic again in 1875, also in Decatur. In OTCO.72, he is listed as a wagonmaker and blacksmith, a business in which his son William H. Darling (a carpenter in Decatur in 1870) was his partner. The Sampson/Davenport 1870 state directory lists him as a carpenter, but *Boyd's Binghamton City and Susquehanna Railroad Directory* of 1871 lists him (by mistake?) as a "mason, builder." By 1875 young William had moved to Roseboom (as a carpenter) and by 1892 to Maryland (as a mechanic).

15. HOMER DENTON (1835 – 1897)

Middlefield(?)N: Middlefield—WFD—Maryland—Schoharie Co. (Seward twp.)—WFD

Apparently a Middlefield native—the family of his father, Richard H. Denton, seems to have been very numerous in that township—Homer Denton was living there with his family in 1850. His parents were members of the Westville Baptist church from at least the mid-forties until September 1849, when they left the congregation; perhaps they simply moved elsewhere in Middlefield then, for they show up again in the 1860 census there. Homer himself is gone from local records until 1870, when at age thirty-five he shows up as a carpenter in Westford. OTCO.72 has him resident in Westville, still as a carpenter. In February 1874, he joined the Westville Baptist church by baptism, but in April he left the congregation (by letter). At this point he became a Baptist minister, serving Schenevus from 1876 to 1880 (he published a sketch of the history of the Maryland church in 1878) and then Seward Valley in nearby Schoharie County. In the early 1880s, he was minister at the Westford church, transferring his membership there in June 1883 and leaving around November 1884. He may have remained in the township thereafter (his wife, Marie Ellis, remained a member of the congregation until 1890 or 1891), and in any case he was there in 1892, when once again he shows up as a carpenter. Five years later he died; he was buried in Middlefield.

16. FERNANDO P. DRAPER (WN: b. 1819)

WN: WFD—Maryland—WFD—Worcester

Brother-in-law of cabinetmaker ALDEN CHESTER, SR., in 1850 Draper himself was practicing that craft in Westford. But he was to be a man of many trades. His parents, Sylvester Draper and Sukey Bigelow, farmed an acreage northeast of the village near Draper Hill, a farm which in later years belonged to James Holmes's sister Lucy and her husband, William Marks, and then to his sister Sarah and her husband, WILL WRIGHT. Fernando himself seems never to have farmed in any large way. He did own ten acres of land from the late 1860s on, and he did run a produce-buying agency in Westford in the 1870s, but these were merely phases of an economic life that included the following: running a hardware store in the village of Maryland prior to 1868 (when he sold it to his brother Lysander); keeping, at about the same time, a Westford hotel that served the students at the Westford Literary Institute, a school which was much connected with the Draper family and which Fernando himself may have been running in 1866; serving as agent for the sale of Mason and Hamlin cabinet organs and Boardman and Gray, Marshall and Traver, and Hayes and Rider pianos in the mid-sixties; and working as a painter in Westford in 1855 and as a mechanic there in 1860. He served Westford as a railroad commissioner in the late 1860s and at the start of the next decade, but otherwise he seems not to have held elective office. By the later 1870s, he was still living in Westford village with his wife, Lucy Preston, serving then as an agent for Albany wool-dealers. Then he seems to have left for Worcester. In 1883 he filed a mechanic's lien against a Worcester resident for whom he had supplied foundation stone; he was himself filed against for a similar reason, but on a different job in Worcester, the same year. Presumably he stayed in Worcester for a longer period, although he has proven hard to trace after the early 1880s.

17. EZRA DUTTON (1825 – 1852)

1850 1855 1860 1865 1870 1875 1880 1890 1900

|....|....|....|....|....|....|....|....|....|....|..

——▪d

New York State N: ?—WFD

Ezra Dutton joined the Westville Baptist church in April 1848 by baptism, on the same day as William Dutton, who must have been his brother, and just a short time before several other members of their family: their sister Lydia, their brother Andrew and his wife, Abigail, and their brother Stephen. But where they suddenly came from—this little family group —has not been determined. They are the only Duttons recorded as members of the church in Westville, and they seem to have left few prior traces in Middlefield, so it would appear that they were not natives of that township. It is certain, too, that they were not Westford natives. Ezra, in any case, called himself a carpenter in the 1850 Middlefield census. His brother William was a laborer in Middlefield then, and from the later 1850s until his death in 1897 would serve as clerk of the Westville congregation. But Ezra was not as fortunate; in November 1852, at the age of twenty-seven, he died.

18. EUGENE ECKLER (b. ca. 1866)

1850 1855 1860 1865 1870 1875 1880 1890 1900
|....|....|....|....|....|....|....|....|....|....|..

Otsego Co. N: ?—Roseboom—?—WFD—?

Nine-year-old Eugene Eckler, living in Roseboom in 1875 with his parents, farmers James Eckler and Catherine Galer, had many important ties with the craft circles of Westford. His mother was the grand-niece of carpenter ANDREW SNYDER (and thus was kin to Snyder's many relatives in the craft—CHANCELLOR, JACOB, and ERASMUS, among others); she also was the sister of carpenter JACOB GALER (and the aunt of his son JAMES) and the aunt of the three HALL brothers. But Eugene appears not to have capitalized on these ties as much as he might. In 1880, by which point his family had moved to Westford, fourteen-year-old Eugene was not living with them. Twelve years later, he showed up again and was working as a carpenter. After that, he was gone from the township.

19. JACOB GALER (1812 – 1875)

1850 1855 1860 1865 1870 1875 1880 1890 1900
|....|....|....|....|....|....|....|....|....|....|..
——d

Middlefield N: Middlefield—Cherry Valley—WFD

Jacob Galer (Gaylord) was the nephew of Westford carpenters ANDREW and JACOB SNYDER but "grew up a shoemaker" (BR.93: 279). A native of Middlefield, where the family of his father, Martin, was numerous and several of the Snyders also resided, Jacob moved north to Cherry Valley, apparently sometime in the 1830s. He may have moved back to Middlefield subsequently; he shows up there in the census in 1840 as engaged in manufacturing and trades. But in the 1850 census he first shows up in Cherry Valley, at age thirty-seven, as a carpenter. His apparent move to Roseboom in 1855 probably was not a move at all—that township was erected from Cherry Valley in 1854—although it is possible that Jacob and his wife, Margaret Brown, and their children moved around 1849 from an area that was to remain part of Cherry Valley into what was to become Roseboom. In any event, through the late 1860s, he was to remain in the new township, still as a carpenter. At that point he moved south to Westford, in company with his son JAMES B. GALER, showing up there as a farmer in 1870 and 1875. In the latter year he was assessed taxes on fifty acres of land worth $350. Also in that year, two days before Christmas, he died. Although never listed as a practicing carpenter in Westford proper, he has been included among the sixty-one carpenters credited to the township because of his strong prior association with the craft and the active presence of his son there at the same time.

20. JAMES B. GALER (b. 1838)

1850 1855 1860 1865 1870 1875 1880 1890 1900

Cherry Valley N: Cherry Valley—WFD—Oneonta

James B. Galer was born in Cherry Valley in 1838, the son of shoemaker-turned-carpenter JACOB GALER, and through him related to a host of Westford carpenters (see EUGENE ECKLER). In 1855, at age seventeen, he already was listed as a carpenter himself in the new town of Roseboom, which had been cut off from Cherry Valley in 1854. He was living with his father and mother then as well as in 1860, when he was still a carpenter; around 1870, he moved to Westford with his parents and his own family (he had married Charlotte Reynolds in Clarksville—i.e., Middlefield village—in 1864). A carpenter in the census that year, in OTCO.72 he is described as a carpenter, farmer, and hops grower. He left Westford around 1873, moving to the boomtown of Oneonta. There he shows up as a carpenter in 1875 and 1892. Calling him "one of the most prominent business men of West Oneonta," BR.93 refers to him as "an undertaker and a mechanic" as well as a "carpenter and joiner" and says that he learned the trade of "a wood mechanic" in Cherry Valley (p. 279).

21. HIRAM GRAVES (b. ca. 1790)

1850 1855 1860 1865 1870 1875 1880 1890 1900

Massachusetts N: Massachusetts—WFD—Oneonta—Laurens

Variously reported as a native of Vermont or Massachusetts (or of some unknown place [1850]), Hiram Graves resided in Westford in 1840, when he was listed as engaged in manufacturing and trades. He has been included among the carpenters of Westford because every subsequent census listing that has been located shows him as engaged in the craft. He was in Oneonta in 1850; had just moved to Laurens a few months before the enumeration was taken there in 1855; was still in that town in 1860; but had moved back to Oneonta by 1865. By then, he was in his middle or late seventies but still gave carpentry as his occupation.

22. HIRAM GRAVES the younger (b. ca. 1848)

1850 1855 1860 1865 1870 1875 1880 1890 1900

New York State N: ?—WFD—?

Very little is known about Hiram Graves the younger. Apparently he was not related to the other man of that name who lived in Westford in the 1840s: that older man had a youngish wife and two children, aged eight and twelve, in 1850, but young Hiram (who would have been about two then) was not listed in his household in Oneonta. Basically, all that is certain is that the younger man was a New York native and that for a brief period around 1870 he was practicing carpentry in Westford. In that year, as a lien he filed in July indicates, in fact he worked as far away as Maryland, repairing a story-and-a-half house there in his capacity as carpenter and joiner.

23. WILLIAM H. GROFF (b. ca. 1819)

1850 1855 1860 1865 1870 1875 1880 1890 1900

Rensselaer Co. N: Rensselaer Co.—WFD—Minneapolis, MN

By 1840, the Rensselaer County native William H. Groff had moved to Westford, where he apparently was living with his kinsman, farmer Nathaniel Groff. Soon afterward, he married Sarah Jane Babcock, the daughter of Westford carpenter RICHARDSON BABCOCK, and it may have been Babcock who taught him the craft. By 1850, the younger man had begun practicing as a carpenter in the township on his own, and in 1855 he still was active. He already had begun to purchase farmland by the late 1850s (he was assessed taxes on 100 acres, worth $1,800, in 1858), and in 1860 he declared himself a farmer. His holdings increased to 123 acres by 1863, and he was once more a farmer in 1865, but probably by that year he had sold his farm and was living in the village proper. Perhaps he was getting ready for what was to happen next, for in 1867 or 1868 he and his family left Westford for Minneapolis, where he hooked up with other kin. In that city, he soon picked up his old craft. Within a few years, however, he had become a millwright—a more demanding trade—and, ultimately, a contractor.

24. NATHAN G. HALL (b. ca. 1828)

1850 1855 1860 1865 1870 1875 1880 1890 1900

Otsego Co. N: ?—WFD—Laurens

Son of Elnathan Hall and Eleanor Galer, brother of STEPHEN and WILLARD HALL (see EUGENE ECKLER for other kin), Nathan was the first of his widowed mother Eleanor's three children to show up in Westford. He was there as early as 1853, part of the great crop of 1855 carpenters, and he remained active through the early part of the following decade. He seems to have done well: he owned $600 worth of land already in 1855 and $1,200 in 1860. But he was not in the assessment books of 1862 and 1863, so he may already have left the township then for Laurens, where he shows up as a joiner in the 1865 census. But he may have tarried longer in Westford after selling his real estate; having joined the Westville Baptist church in 1859, he did not leave it (by letter of transfer) until January 1864. The 1868 county atlas locates him as Rev. N. G. Hall in Mt. Vision, a village in Laurens. The census in 1870 shows him as a carpenter again, and as owning $3,000 in real estate; the state directory of that year and OTCO.72 confirm his occupation (OTCO.72 has him as a carpenter and builder) and list him as still in Mt. Vision. In 1892, still listed in Laurens, he was a builder, pure and simple.

25. STEPHEN HALL (b. ca. 1830)

1850 1855 1860 1865 1870 1875 1880 1890 1900

Otsego Co. N: ?—WFD—?

Stephen Hall, like his older brother NATHAN, was connected by kinship ties with several other Westford carpenters. And the tie with Nathan himself was important. Stephen followed his brother to Westford:

Nathan came in 1853 and Stephen in 1855, when he was boarding with Nathan there. But soon their paths diverged. Whereas Nathan seems to have stayed with house construction until he left the township by 1864, Stephen had become a wagonmaker by 1860 and remained there after Nathan's departure. The 1860 census indicates that he made only two wagons the previous year (valued at $150); he also did repairs and produced other articles (worth another $350). He was still there in 1865 and still a wagonmaker. By 1870, however, Stephen had left Westford for an unknown destination.

26. WILLARD HALL (ca. 1832 – 1878)

1850 1855 1860 1865 1870 1875 1880 1890 1900
|. . . . |. . . . |. . . . |. . . . |. . . . |. . . . |. . . . |. . . . |. . . . |. . . . |. . . . |. .
d

Otsego Co. N: ?—Hartwick—Middlefield—WFD—Laurens

The youngest of the three Hall brothers, Willard was living as a farmer (evidently a farm laborer) with a family in the township of Hartwick in 1850. Ten years later, he was a carpenter in Middlefield, probably near the border village of Westville, where his brother NATHAN (his employer then?) seems to have been living at the time. The association with Nathan appears to have held over time, but in 1865, when Nathan first showed up in Laurens, Willard (like his other brother, STEPHEN) was living in Westford township, still as a carpenter. (His apparent shift from Middlefield to Westford township that year, by the way, may have entailed nothing more than a move across the street in the border village of Westville.) In 1875, when Stephen apparently had left the region, Willard was back with Nathan. Active himself then in Laurens, Willard was to die in three years.

27. GEORGE HOLDRIDGE (ca. 1833 – 1859)

1850 1855 1860 1865 1870 1875 1880 1890 1900
|. . . . |. . . . |. . . . |. . . . |. . . . |. . . . |. . . . |. . . . |. . . . |. . . . |. . . . |. .
d

Otsego Co. N: ?—Maryland—WFD

A seventeen-year-old laborer living with his parents (John and Harriet Snyder Holdridge) in Schenevus in 1850, George had moved to Westford by 1855. There he was living with a cousin, farmer James Southard, and was working as a carpenter. Westford was hardly unknown ground to him, however. His father, John, was the son of Thomas and

Hannah Holdridge, who lived in Westford from 1830 on. His mother, probably from Middlefield, was not a Westford SNYDER, but she had joined the Westford Baptist church in 1838, along with several relatives, by a letter from the congregation in Waterville (i.e., Westville). His father, who had joined by baptism in 1842, was still active in the congregation as late as 1847. George himself apparently never joined the congregation, and in any case he died young in April 1859, when he was only twenty-six. Later his parents and his brother William were the proprietors of the new hotel near the railroad in Schenevus, the Holdridge House. There—so the Cash Account in his diary dutifully records—Holmes spent twenty-five cents in late January 1869 when en route to Colliersville via the train.

28. JAMES C. HOLMES (WN: 1830 – 1895)

WN: WFD—Delaware Co. (Delhi twp.)—WFD

The sixth child and fourth son of Calvin Holmes and Mary Kelso, James Calvin Holmes grew up on the family's Cold Spring Farm north of Westford village. In 1850, he was living with Dr. Calvin Howard in the village of Delhi in Delaware County as a student of medicine. He was gone from Westford for about five years altogether, but by 1855 he had returned, been married, and declared himself a carpenter. He and his wife, Mary Shelland, were living then with farmer Chester Wright, brother of carpenter MILTON WRIGHT, who probably trained Holmes's brother JOHN and Holmes himself. By 1860, James and Mary had purchased from his stepmother, Anna Howe Holmes, a house and five acres of land adjoining the village (the home is visible in fig. 2). Holmes was still a carpenter, both then and in 1865; but the purchase of almost eighty acres of land in 1868 probably signaled an intent to enter farming in a fairly serious way. The transition came in the next decade: Holmes still claimed in 1870 that he was merely a carpenter; in 1875 he was a farmer and carpenter; and in 1880 he was simply a farmer. Following the death of his wife in 1878, he married (in 1881) Daphnea V. Cook Bentley, widow of STEPHEN BENTLEY. Very active from the 1850s in the Westford Baptist church, he served it as clerk from 1862 to 1869 and in many other capacities over the years: as a trustee, the treasurer, the superintendent of the "Sabbath School," a member of various committees, and often a delegate to meetings of the Worcester Baptist Assocation, the local denominational body. Holmes also was elected town clerk in 1861 and held various minor offices connected with elections over the following fifteen years. To that extent his life in Westford was a modest success. But his economic condition must have worsened after around 1875, perhaps because craft work was becoming scarce. His sister Clarissa took title to the land where his house and barn stood in 1878, probably as security for old loans (he was in her debt as early as 1866) or for new ones; she was paid off in 1881 by Holmes's new wife, but Daphnea at the same time took title to those parcels and the forty-five-acre remainder of the farm. Holmes was never again to own land in Westford, yet he nonetheless kept tending the small farm until his death—in the barn there—in 1895.

29. JOHN K. HOLMES (WN: 1827 – 1907)

1850 1855 1860 1865 1870 1875 1880 1890 1900

WN: WFD

The only one of Westford's carpenters who seems never to have left the township, John Kelso Holmes was one of JAMES C. HOLMES's older brothers. He almost certainly was trained in carpentry by MILTON WRIGHT, with whom—already a carpenter himself—he was living in 1850. The ties between the two men were multiple: John Holmes married Wright's half-sister Louisa in 1851 and in 1855 was partner with Wright's half-brother HENRY WRIGHT in a sash-and-blind factory. (Apparently this was the same factory, at the north end of the village, that ALDEN CHESTER was running soon afterward and that his son HORACE was running in 1860.) In 1855, the Holmes-Wright operation marked an investment of $650; it used $500 worth of pine logs per year to produce, by waterpower, $1,000 worth of sash and blinds. Only the two owners worked in the factory, receiving an average monthly wage of $30 each. By 1860, when Horace Chester was running the operation, John Holmes was listed simply as a carpenter (Henry and WARREN WRIGHT, seemingly employed then by young Chester, were sash-and-blind makers). By 1858, Holmes was listed as the owner of a lot in Westford village worth $250, a property he seems to have kept for much of the rest of his life. He was a mechanic in 1865 and—perhaps there had been no real change of occupation here—a wagonmaker in 1875. Unaccountably absent from the U.S. Census for Westford in 1880, he was counted there then in a special state enumeration which, however, did not record occupations. In the next surviving census, the state count of 1892, he was listed as a mechanic, but thereafter his fortunes appeared to have declined: in 1900 he was merely a day laborer.

30. HARRISON HUBBARD (WN: 1840 – 1908)

1850 1855 1860 1865 1870 1875 1880 1890 1900

WN: WFD—?—WFD

Born in Westford, where his parents Harrison and Eliza Hubbard had been farming from the late 1830s, young Harrison left the township with them by 1855. But the break with Westford was not permanent. His father and mother were buried in nearby Middlefield in 1857 and 1866, respectively. And the young Harrison's brother David, after being elsewhere, returned to Westford by the early 1870s, when he was farming fifty-three rented acres. Harrison himself also seems to have been nearby then: OTCO.72 lists him as farming one hundred acres in Middlefield in 1870. In any case, by 1892 young Harrison (then, at fifty-one, no longer "young") had come back to Westford as a carpenter. In 1900 he was still practicing there, probably in the village of Westville. At age sixty, he reported that he had worked only four months out of the previous year, but he remained nominally active at least until 1905, when he and his wife, Louisa Reynolds, were still residing in the township.

31. GILES LAMPMAN (b. ca. 1808)

1850 1855 1860 1865 1870 1875 1880 1890 1900

Schoharie Co. N: ?—WFD—Unadilla

Giles Lampman, born in nearby Schoharie County, first arrived in Westford in 1847 or 1848. He was a forty-three-year-old carpenter there in 1850, living in Westford village with his wife, Louisa, and their two children and his mother, Catherine, a native of Troy, New York. Two more children were added to the household over the next decade, as Giles continued to practice his craft. The Lampmans remained in Westford as late as 1867 or 1868, when the new county atlas showed him as still the owner of a house in the village proper; but by 1870 they had moved to Unadilla township. There Lampman was still a carpenter in 1870 and 1875, although OTCO.72 describes him as a farmer as well as a mechanic. He was then residing in the village of Wells Bridge, "a flourishing little village," as Hurd's county *History* put it in 1878, "rapidly becoming an important station on the Albany and Susquehanna railroad" (p. 340). Lampman thus seems to be another of those Westford carpenters who left the township for brisker economic opportunities elsewhere.

32. DANIEL LEONARD, JR. (b. ca. 1817)

1850 1855 1860 1865 1870 1875 1880 1890 1900

|....|....|....|....|....|....|....|....|....|....|..

Schoharie Co. N: ?—Cherry Valley—WFD—Middlefield—Otsego

Born in Schoharie County, Leonard was farming in Cherry Valley by 1850, but when he moved to Westford in 1855 it was as a carpenter. He owned land there then; in 1860, still a carpenter, he owned none. He and his wife, Jelanie (there are many bemused and amusing spellings), and their four children moved to Middlefield by 1865. There the "Products of Industry" schedule in the census of 1870 shows him as fully employed for the past twelve months, with a capital investment of $200, an expense for materials of $210, and a production of $500. By 1875, the family had relocated to Otsego township, where they owned a house worth $2,500.

33. CHARLES LOOMIS (1819 – 1912)

1850 1855 1860 1865 1870 1875 1880 1890 1900

|....|....|....|....|....|....|....|....|....|....|..

Egremont, MA N: Egremont—Middlefield—Maryland(?)—WFD—Lenox, MA

Only briefly associated with Westford, Charles Loomis arrived there in April or May 1855 and was gone within a few years. He had moved to the region from his native Massachusetts sometime in the 1840s, locating first in Middlefield and then perhaps moving to Maryland. He was married to a Westford woman, Emily H. Badeau, in 1847, and in 1850 the census showed them as living in Middlefield in a house they owned worth $600. Since Emily switched her church membership from Westford village to Westville in 1851, they must have been close to the township border then. Charles himself joined the same Westville congregation by baptism in 1853. By 1855, however, the family was living with Emily's parents in Westford proper. It is possible that they had suffered some kind of reversal (Charles no longer claimed to own any land, and in mid-1854, their five-year-old daughter Ann had died); whatever the case, things turned even worse the following fall, when Emily, just thirty-one, died and was buried in her family's cemetery on Badeau Hill. For unclear reasons, Charles was excluded from the Westville church two years later, and by 1860 he had left the township. His daughter Emma was living in Westford with her Badeau grandparents in 1860, with her mother's brother in 1865, and with the William Skinner family in 1870. But soon she joined her father, who had moved back to Massachusetts, and in 1880 she married carpenter John H. Root in Lenox. Charles had two more wives before his death in 1912; he was buried with Emily Badeau and little Ann back in Westford.

34. PHILIP MORRIS (b. ca. 1790)

1850 1855 1860 1865 1870 1875 1880 1890 1900

New Jersey N: New Jersey—WFD—?

New Jersey native Philip Morris came to Westford around 1837 or 1838, and in the 1840 census he was described as a farmer. He bore the same label in 1850, when he was fifty-nine, but five years later (when his age was given as sixty-six), he was a carpenter. After that, he vanished from the township.

35. JOHN PANK (b. ca. 1840)

1850 1855 1860 1865 1870 1875 1880 1890 1900

Cherry Valley N: Cherry Valley—WFD—?

Robert and Mary Ann Pank, born in England, had come to Cherry Valley by 1838. They farmed there until just after 1855 and then moved to Westford. The 1860 Westford census shows them still farming, but their son John, listed with them then, was a nineteen-year-old carpenter. He had left the township before his brother WILLIAM, not with the family in 1855, briefly rejoined them in their new home.

36. WILLIAM PANK (b. ca. 1839)

1850 1855 1860 1865 1870 1875 1880 1890 1900

|....|....|....|....|....|....|....|....|....|....|..

Cherry Valley N: Cherry Valley—WFD—?

William, JOHN PANK's brother, was absent from his family in 1850 and 1855 but showed up with them following their move to Westford. Just twenty-one in 1860, he was then a laborer. He was not counted in the township in 1865, but OTCO.72 lists him there once more, as a carpenter and farmer.

37. ANDREW J. PETTE (WN: b. ca. 1826)

1850 1855 1860 1865 1870 1875 1880 1890 1900

|....|....|....|....|....|....|....|....|....|....|..

WN: WFD—Roseboom—?

Aaron Pette, a native of New Jersey, had come to Westford around 1810 and married his wife, Betsey, there in the 1820s. They moved with their sons, Andrew and GEORGE, to Worcester around 1830, but they came back to Westford within four or five years. There they were farmers, as was Andrew in 1850, when he first was listed with an occupation. By 1855, however, he had become a carpenter. Five years later, he had moved to Roseboom, where he was practicing the same craft; then he dropped from the local record.

38. GEORGE C. PETTE (WN: b. ca. 1828)

1850 1855 1860 1865 1870 1875 1880 1890 1900

|....|....|....|....|....|....|....|....|....|....|..

WN: WFD—?

Like his brother ANDREW, George Pette first showed up as a farmer in Westford, in 1850, but five years later he had become part of the extraordinary 1855 crop of twenty-two carpenters. By 1860, he had left the township, apparently for good.

39. MARTIN PIERCE (WN?: 1812 – 1895)

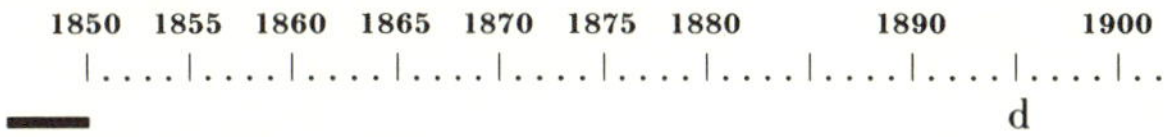

WN(?): ?—WFD—Roseboom

Martin Pierce's parents, Sheldon and Sally Pierce, were farming in Westford by 1820, but when Martin himself first was listed in the Westford census—in 1840—he was already engaged in manufacturing and trades. By 1850, he had moved to Cherry Valley, where he was then a thirty-nine-year-old laborer. He continued to live for the rest of his life in the part of that township that was erected into Roseboom in 1854. By 1860 he was calling himself a carpenter, a label that he continued to claim there into the 1870s. Because of this strong association with the craft later in his life, he is counted as one of the sixty-one Westford carpenters even though he never was identified as one there.

40. JULIUS E. PRESTON (WN: b. ca. 1846)

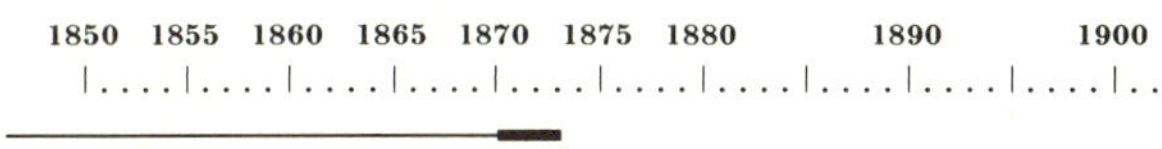

WN: WFD—Maryland—?

Julius Preston's father, Leonard, was a farmer in Westford for most of his life, although in 1870—at the age of fifty—he suddenly was listed as a maker of boots and shoes. Julius was himself an eighteen-year-old farmer in 1865; in 1870, he was a carpenter. Two years later, he was living and working in the village of Schenevus as a telegraph operator, one of the new careers that the railroad had brought in. His parents, also Westford natives, likewise had left Westford around the same time. None of them ever seems to have returned.

41. FRANK RICKARD (b. ca. 1854)

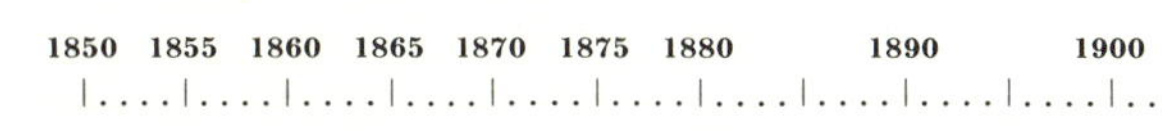

[Unknown]: ?—WFD—?

Briefly resident in Westford as a carpenter in the early 1890s, Rickard was gone by 1900. His parents, Elias and Marion, were living with him and his own wife and two children there in 1892.

42. HEZEKIAH SKINNER (b. 1831)

Schoharie Co. N: Schoharie Co.—WFD—Delaware Co. (Tompkins twp.)—WFD—Worcester

Son of John Skinner and Anna Smith, who settled in Westford in the 1830s, Hezekiah was tied by marriage to both MILTON WRIGHT and to JAMES and JOHN HOLMES. He probably learned carpentry from one of them before he left Westford around 1850 to work for his sister Lucinda's husband, Jacob H. Wright, in Delaware County. Jacob Wright (no relation to Milton Wright) was a housebuilder there in 1850; Skinner, living next door with two other young men, was a carpenter. After Wright's death around 1854, Skinner returned to Westford and was listed as a carpenter there in 1855 and a farmer five years later. When he showed up in Worcester in 1865, it was as a carpenter once more.

43. ANDREW/ANDREWES/ANDRUS SNYDER (1796 – 1889)

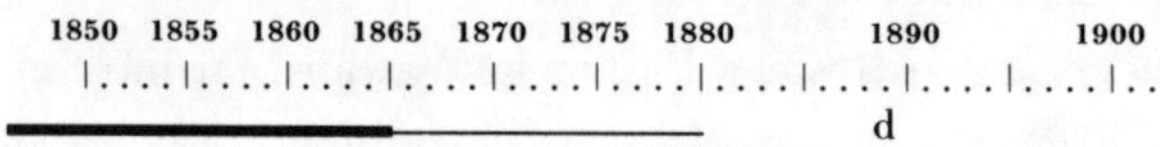

Rensselaer Co. N: Rensselaer Co.—WFD—?

Andrew Snyder came to Westford around 1805 with his widowed mother, Elizabeth, and her many other children. Already in 1820, he was listed in the census as engaged in manufacturing and trades there, a listing repeated in 1840; the surviving daybook of Dr. Elhanan Jackson shows him performing various woodworking jobs in the mid-1830s, so presumably his trade during this period in fact was carpentry. Indeed, connected by kinship ties to many of the other carpenters in the township—all the other SNYDERS, plus the HALLS, the GALERS, and others—Andrew was at the center of craft activity there for much of his life. The censuses for 1850, 1855, and 1860 show him as a carpenter, but thereafter (until 1875) he appears as a farmer. Gone from the township in 1880, he apparently had moved elsewhere to stay with kin—perhaps his brother John, also gone by then, or his son (by his second wife, Emmeline Allen) Theodore Allen Snyder. At his death in 1889, he was buried in Westford.

44. CHANCELLOR/CHAUNCELAER SNYDER (1824 – 1895)

1850 1855 1860 1865 1870 1875 1880 1890 1900

d

Middlefield N: Middlefield—WFD—Middlefield

The son of JACOB and nephew of ANDREW SNYDER, Chancellor Snyder was a carpenter living with his father (also a carpenter) and his brother Jerome (a blacksmith) in Westford in 1850. They had been in Middlefield in 1840, and it seems likely that Chancellor had been born there in the mid-1820s. His father and brother remained in Westford, but Chancellor had moved back to Middlefield by 1855 and was practicing there both then and in 1860. The county atlas published at the end of that decade shows his house just north of Westville and just west of the township line. Apparently he prospered in this vicinity: the atlas seems to show him as the owner of two other houses, one across the road and one in Westford township, and the 1870 census listing of him (still as a carpenter) states his worth as $12,500, a sum quite large by local standards. OTCO.72 may explain this figure by its description of him as a "carpenter and builder, undertaker and farmer." Probably the cabinet shop shown in the 1868 atlas was his (cabinetmakers often doubled as undertakers, of course). He later seems to have confined his work to undertaking, his occupation in the 1892 state census.

45. ERASMUS SNYDER (WN: 1822 – 1898)

1850 1855 1860 1865 1870 1875 1880 1890 1900

d

WN: WFD—Middlefield—Cherry Valley—WFD—Middlefield

The son of ANDREW SNYDER's twin brother, Christopher, a farmer, Erasmus Snyder must have learned the craft in Westford, perhaps from his uncle. He left the township in 1849 and in 1850, living in Middlefield, was already listed as a carpenter. He was still practicing there five years later. In 1860, however, he and his wife, Julia Hall, were living in Cherry Valley as domestics in the house of her mother. They returned to Westford by 1866, when he was assessed taxes on property there, or by 1867, when she joined the Baptist church by letter of transfer. The 1868 county atlas shows an E. Snyder as owning a house at the south end of the village. Holmes refers to Snyder in his 1869 diary, and Snyder shows up in the census the following year as a carpenter, worth $2,800, living near Holmes's brother JOHN. But this new association with his native township was not to last. Perhaps due to some economic misfortune, Snyder owned no land when he was listed in Middlefield again in 1875 (the Baptist church records in Westford say, on February 28, 1874, that he "removed to Middlefield"). He was still practicing his craft there, as he was when last listed in 1892, six years before his death.

46. JACOB L. SNYDER (1789 – 1868)

1850 1855 1860 1865 1870 1875 1880 1890 1900

d

Rensselaer Co. N: Rensselaer Co.—WFD—Middlefield—WFD

Brother of ANDREW and father of CHANCELLOR SNYDER, Jacob L. Snyder came to Westford with his widowed mother around 1805. He relocated to Middlefield prior to 1840 but was back in Westford again by 1850. He practiced carpentry there for the next few years, living with his son Jerome, a blacksmith. In 1855, he and Jerome—carpenter and blacksmith though they remained—in fact were running a water-powered shingle factory. That operation, a $600 investment, used 500 logs a year (valued at $600) to turn out 14,000 bundles of shingles worth $1,600. Three men worked there, averaging a monthly wage of $30. Jacob Snyder declared himself a shingler-cutter in the 1860 census, but by 1865 he was once again a carpenter.

47. ELISHA SPAFFORD (b. ca. 1829)

1850 1855 1860 1865 1870 1875 1880 1890 1900

WN: WFD—?—Maryland—Worcester—?

Elijah Spafford, father of carpenter Elisha Spafford, was one of the founding members of the Westford Baptist church in 1825 and remained an important member—indeed, an elder—in that institution for many years. Since Elijah became a minister who served a number of local congregations, however, his son Elisha lived only intermittently in his native Westford in his youth. He first showed up on his own in the township in 1850, when (aged twenty-one) he was listed as a farmer. Five years later he was a carpenter living near AARON W. BENTLEY and PETER BROWN. He was still there in 1860 but left during the next decade, showing up in Maryland as a farmer in 1870. OTCO.72 showed him in the same occupation but living in Worcester (served by the East Worcester post office). He was in that township again in 1875 and had by then resumed his craft. A mechanic's lien filed by him in 1880, as an assignee for a Milford man, suggests that he was active at that time as well.

48. EZRA THURBER (1792 – 1876)

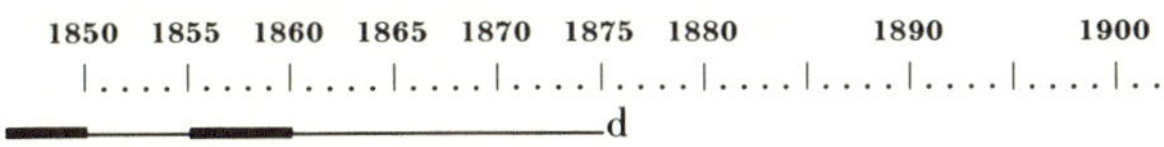

Vermont N: Vermont—WFD

Ezra Thurber was, along with WILLIAM BENTLEY and ANDREW SNYDER, one of the three most important carpenters of early Westford. He came into the vicinity from Vermont (although one census listing says Massachusetts, and in 1900 his son reported Rhode Island) between 1810 and 1820, and by the latter year he already was listed as engaged in manufacturing and trades. He was still in Westford in 1830 and was active throughout that decade; the daybook of Dr. Elhanan Jackson, for instance, shows him at work on a shed in 1836. In 1840, he was once more listed in manufacturing and trades. (The second person in his household shown as active then may have been his eleven-year-old son OLIVER, although that is unclear.) Ezra called himself a farmer in 1850, when he owned $1,200 worth of real estate; but he claimed only $200 worth in 1855, when he was again a carpenter. Thereafter, following the purchase of some fifty-seven acres of land by 1858, he was to farm until the early 1870s. His farm was worth $2,500 in 1870, but a year before his death (which occurred in 1876), he was shown as owning no land and—as a retired farmer—as having no occupation. Oliver, living with Ezra and with his mother, Eliza, was a farmer in 1870 and 1875, so he may have taken over the farm from his parents.

49. OLIVER THURBER (WN: 1830 – 1916)

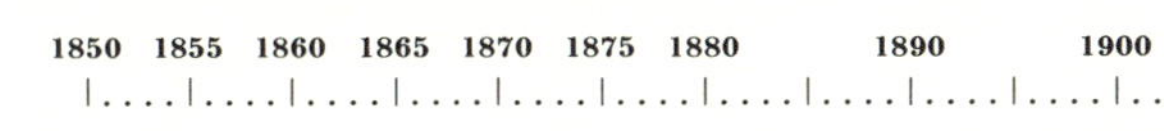

WN: WFD—Maryland—WFD

Although born and raised in Westford, where his father EZRA was long an important carpenter, Oliver Thurber shows up in the craft for the first time in nearby Maryland. By 1855, he had been there four years and was then the owner of land. He came back to Westford by 1870, when he was farming with his parents. Ezra remained the head of the household in 1875, but he was retired from farming by then. It was Oliver, still a farmer then, and already listed in OTCO.72 as owning ninety-seven acres, who apparently held the family farm. Oliver persisted as a farmer in 1880, but OTCO.72 had added "carpenter" to that description; in 1892 he was simply a carpenter once more. The 1900 census credited him with no farm—he owned a house, but it was mortgaged—and noted that he had been employed as a carpenter for eight months in the previous year, certainly impressive for a man near seventy. One job, begun in April 1900, was mentioned then in the Schenevus paper—"extensive repairs" on the barns of Westford farmer Amasa Skinner. In 1905, at seventy-six, he once again was a carpenter, but in 1915 he was an eighty-five-year-old widower, living alone, without a declared occupation. His second wife—or perhaps third—Abigail Snyder (ANDREW SNYDER's niece) had died that year, and Oliver was to follow her in 1916.

50. DAVID TILTON (b. ca. 1814)

1850 1855 1860 1865 1870 1875 1880 1890 1900

Maryland (State) N: Maryland (State)—WFD—?

A cabinetmaker who in 1850 claimed he had been born in Maryland, David Tilton had no clear kinship ties that might have led him to Westford, and when he left the vicinity soon afterward he left no apparent traces. A few people with his family name have been found in Worcester, but he cannot be placed among them by means of any available records. Given the spareness of pre-1850 census records, it also has not been possible to trace him in the state of Maryland.

51. WILLIAM WALES (1822 – 1897)

1850 1855 1860 1865 1870 1875 1880 1890 1900

d

England N: England—WFD—Cherry Valley—Middlefield—WFD

Englishman William Wales had declared his intention to become a U.S. citizen in Philadelphia in 1848 and had been naturalized in 1851

(P. William Filby, ed., *Philadelphia Naturalization Records* [Detroit: Gale Research, 1982], p. 673), although he may have come to this country well before the late 1840s. A farmer resident in Westford when he became a citizen, in 1854 he had moved with his wife, Elvira, and their children to Cherry Valley, where he was to practice carpentry. In this he resembled three men who very likely were his brothers—Edward, Robert, and Philip Wales; all were English natives of the same generation, and at various times all were active in Cherry Valley. William Wales had shifted his domicile to Middlefield by 1860. He was still a carpenter at that time, but perhaps the family was farming then, too. In 1858 Wales was assessed taxes in Westford on ninety-seven acres there; the family may have lived close to the Westford border, or their farm may have straddled the line. If either of these hunches is correct, the family's move back to Westford before the 1865 census was taken may not have involved much of a change. But an increase in the number of acres for which Wales was taxed, as recorded first in 1868, and the fact that in 1869 he became a member of the Baptist church in Westford village may argue for his residence nearer the township's center by 1865. The 1868 county atlas likewise shows him living on the Middlefield road about two miles northwest of Westford village and not at all close to the township line. However these matters stood, the family's future ties with the township were to be lasting. Elvira Wales died there in 1879; in the 1880s, William married again and was to farm in Westford for the rest of his life. Although never technically listed as active in Westford records, he has been included among the sixty-one Westford carpenters because of his long prior association with the craft nearby.

52. DELOS WALLROD (b. ca. 1824)

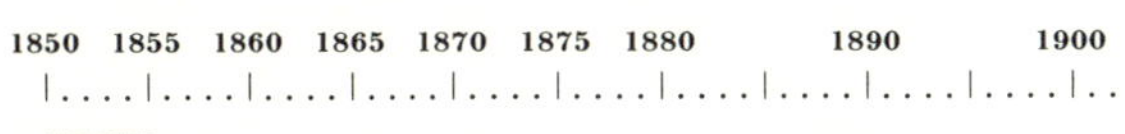

Unknown: ?—WFD—?

A twenty-six-year-old carpenter of unknown origin listed in Westford in 1850 with a wife and two children, Wallrod had left the township by 1855. He may possibly have been from Cooperstown, where a number of "Walradts" (a more conservative spelling of the German name) practiced carpentry during the midcentury. But in that case Delos Wallrod would have been listed twice in the 1850 census, since the Cooperstown schedule that year itself contains "Delos Walradt," a twenty-five-year-old carpenter living with a woman, Betsey Walradt, who could have been his mother. An outside possibility is that Wallrod's wife, Emma, and children were in Westford, while he was working then in Cooperstown; he might well have been counted with his own family, as was reasonable enough, but also listed where he was then residing. The Westford listing places his family in the house of Lary Way, a fifty-one-year-old laborer who could have been Emma "Wallrod's" father. Of unknown birthplace himself, Way was never present in Westford in any earlier or later census or in other records. If the Westford and Cooperstown men were one and the same, then Wallrod's persistence in the county would have been higher: he was still in Cooperstown in 1870, and still employed as a carpenter there.

53. GAIUS WARD (b. 1857)

1850 1855 1860 1865 1870 1875 1880 1890 1900

|....|....|....|....|....|....|....|....|....|....|....|..

Otsego Co. N: ?—Oneonta—WFD—?

Gaius Ward first showed up in Westford as a carpenter in 1900, and in 1905 he was still practicing there. In the earlier year, he was living as a boarder with a kinswoman (so it would seem), a forty-five-year-old widow named Minnie Ward. By 1905, still single, he had become a boarder in another house. He first shows up in the census in Oneonta in 1875; then eighteen and without a stated occupation, he was living with his brother Chauncy, a thirty-four-year-old carpenter. By 1900, Chauncy had moved to Westville; the Schenevus paper in March of that year reported his death there after a "long illness of creeping paralysis." He had been, the paper opined, "a first-class carpenter" (*Schenveus Free Press and Davenport Standard*, 3/7/1900) and was to be buried in Oneonta. Chauncy's wife in 1875 was named Ella, so the Minnie of 1900 was probably not the same woman. However, the fact that Minnie's son in 1900 had a last name different from hers (he is listed as Menzo Race, aged eighteen) suggests that Minnie just may have been Chauncy's second wife, and thus Gaius Ward's sister-in-law. If so, she would have been widowed just recently (again?) when the 1900 census was taken.

54. HORACE WEBSTER (b. ca. 1820)

1850 1855 1860 1865 1870 1875 1880 1890 1900

New York State N: ?—WFD—?

Born somewhere in New York, Horace Webster showed up in Westford as a forty-year-old carpenter in 1860. He was living near the wagon-maker W. H. Platner and claimed real estate worth $1,200 and personal property worth $1,000. He married Westford native Amanda L. Preston sometime prior to that year, perhaps somewhere other than Westford, since although she is included in the 1850 census there with her parents, neither she nor Horace appears in Westford in 1855. That he was taxed solely for $800 of personal property in Westford in 1863 may mean that his real estate was located elsewhere, but it could mean that—since the death of Amanda the year before—he had sold his land there in preparing to leave the township. In any case, by 1865 he was gone.

55. NEHEMIAH WHITE (WN: b. ca. 1824)

1850 1855 1860 1865 1870 1875 1880 1890 1900

WN: WFD—?

Nehemiah White may have been the son of Peter White, a Westford farmer who died sometime in the 1840s. In any case, he was living there with his widowed mother, Phebe, and his brother Henry in 1850; both of the sons were farmers. In 1855, when he was listed next to Benjamin White, conceivably another brother, Nehemiah was a carpenter. He had left the township by 1860. He may have been the man of that name who enlisted in the army, from nearby Decatur, in 1864, but no further positive record of him has been found.

56. JUDSON WINANS (b. ca. 1839)

1850 1855 1860 1865 1870 1875 1880 1890 1900

Livingston Co., New York/Michigan N: ?—Michigan—WFD—Oneonta—Worcester

Judson Winans was likely the son of sometime Westford resident William Winans, Jr., who was active in the Westford Baptist church in its early years but left it in 1836. Judson himself probably was born in Livingston County, New York, although one census does give Michigan as his place of birth. In any case, he must have migrated with his family to Michigan at some point, since he and his siblings seem to have been living there when they were orphaned around 1854. Judson and the others had been brought to Westford in 1855 and were living then with their aunt Louisa Winans and her husband, carpenter MILTON WRIGHT. By 1860, Judson was married to Westford native Louise Darling, and they owned their own house. He was employed then as a laborer—an ambiguous term which may mean that he was at work either in carpentry or agriculture. In 1865, he clearly was working as a farmer, although the family owned no farm and had no separate entry of their own in the agricultural census then; indeed, they were living with Louisa's parents. On the other hand, Winans clearly had entered the craft by 1868 at the latest, since he then had begun work on a new house north of Westford village, the house JAMES HOLMES was at work on when he began his diary for 1869. Probably Winans (like Holmes) had been trained by Wright; but when he worked for Holmes in 1869, Holmes clearly was supervising his labor, and Winans—while certainly more skilled than Everett Herdman—appears not to have been a full-fledged carpenter yet. He claimed that label for himself in a business directory which must have been solicited for in 1869. And in 1870 he was listed as a carpenter in the census in Oneonta, the booming railroad village at the southern edge of the county. OTCO.72 repeats the claim, but Winans had no occupation in the Oneonta census in 1875, or in 1892, by which point he had moved with his family to Worcester.

57. HENRY E. WRIGHT (WN: 1827 – 1864)

WN: WFD—Virginia [with army]

Brother of WARREN WRIGHT, half-brother of JOHN and MILTON WRIGHT, uncle of WILLIAM E. WRIGHT, and kin of the HOLMESES, Henry E. Wright was located near the center of an important craft network in midcentury Westford. It must have been from Milton that he received his training, since he was living (as was John Holmes) in Milton's house as a carpenter when the 1850 census was taken. In 1855, when he was back in the house of his father, Pomeroy Wright, a farmer and long an important member of the Westford Baptist church, no occupation was listed for him, but in the "Products of Industry" section of the census, he is listed as running, with John Holmes, a sash-and-blind factory in the village. (For statistics on its output, see JOHN HOLMES.) Late that year—perhaps this move marked the point at which he and Holmes sold the factory to ALDEN CHESTER, SR.—Henry, himself a Westford Baptist, requested a traveling letter or "commendation" from the church, usually a mark that a member was going to leave the township but had no definite destination in mind. It is not clear, however, whether he left the vicinity or, if he did, exactly when he came back. He was assessed a military head tax in 1858 and was next mentioned in the church records in 1859. By 1860, he was working as a sash-and-blind maker, but the factory then was being operated by Chester's son DWIGHT, and Henry must have been simply one of the four men reportedly employed in it. In January 1862, still unmarried, Henry Wright enlisted in the Eighty-first New York Regiment. On June 2, 1864, he was killed in the Battle of Cold Harbor in Virginia.

58. JOHN WRIGHT (WN: 1809 – 1887)

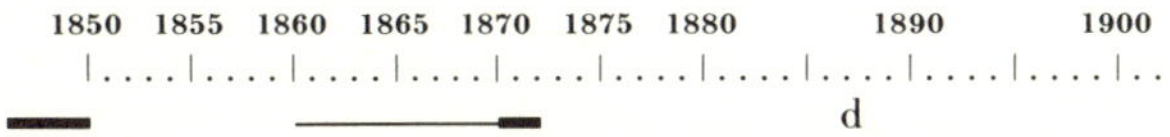

WN: WFD—Worcester—WFD—Maryland

Brother of MILTON WRIGHT and kin of several other Westford woodworkers, John Wright seems to have been gone from the township in the 1830s but was engaged in some craft there as early as 1840. Perhaps he was then a wagonmaker, as he would be later in Worcester (1850) and in Schenevus (OTCO.72). But Wright was a man of several occupations. He ran a hotel in Westford from the later 1850s up to at least 1865, and he himself may have erected the building for it. In 1870, he was a painter in the village, though the fact that he lived next to wagonmaker William H. Platner then may suggest that he really was working with or for Platner. In the following census, while residing in Schenevus, Wright was a painter and glazier. Perhaps he was working for Lane & Hotchkin's sash-and-blind factory at the time. OTCO.75 stressed his variety of talents by listing him not only as a wagonmaker, as already noted, but also as a carpenter and painter.

59. MILTON WRIGHT (WN: 1812 – 1873)

1850 1855 1860 1865 1870 1875 1880 1890 1900

|....|....|....|....|....|....|....|....|....|....|..

d

WN: WFD—Worcester—WFD—Maryland

The center of an important craft network that included the many WRIGHTS active in Westford as well as the HOLMESES and his nephew JUDSON WINANS, Milton Wright began practicing in his native township in the 1830s. He and his wife, Louisa Winans, left the Baptist church there in 1835, however, and relocated in Worcester. They were still in that township in 1840, when their two-year-old child died falling down a set of stairs. Milton was listed as engaged in manufacturing and trades in the 1840 census there, and in 1841 he was a founding member of the new Baptist congregation in the village of Worcester (Louisa joined later that year). But then, in 1843, the Wrights left Worcester and returned to Westford, where Milton was elected Baptist clerk (1844), a position he held for almost two decades—though with a few gaps—until JAMES HOLMES replaced him in 1862. Wright held many other positions in the congregation and was indeed at the heart of the church's life. In 1850, with a family that included future carpenter WILLIAM E. WRIGHT, Milton was working as a carpenter in Westford, and carpenters JOHN HOLMES and HENRY E. WRIGHT were living with him. He was still active in 1855, when future carpenter Jud Winans was in his house. But in the later part of the 1850s Wright seems to have begun accumulating land in order to effect a change from carpentry to farming. He owned two parcels totaling 162 acres in 1858, a large quantity by local standards; the land was valued at $2,300 in the assessment that year, also a high figure. In 1860, at forty-nine, Wright was listed in the census as a farmer, but in 1863 his holdings fell to 80 acres, and in 1866, to a village parcel and an outlot of a mere 19 acres. If there was a real reversal here, at least Wright had his craft to fall back upon: he called himself a carpenter again in 1865. Still active in the Westford church, in 1869 he and his wife asked for letters of dismission from it so that they might move—as their son Will already had—to the village of Schenevus. It must have been the promising economic changes then going on there, as the railroad line was completed, that attracted them, as they did so many others. But the Wrights seem not to have flourished in their new home. Owning no real estate and having a personal estate of only $300 in 1870, Milton was listed as a carpenter once more in the census that year. A mechanic's lien he filed in January of that year shows that he indeed still was working. In OTCO.72 he was listed once again as a carpenter, but soon Milton Wright died, at the age of sixty-one in February 1873, and was buried in the Worcester cemetery.

60. WARREN WRIGHT (WN: b. ca. 1838)

1850 1855 1860 1865 1870 1875 1880 1890 1900

WN: WFD—?

Brother of HENRY E. WRIGHT and half-brother of MILTON WRIGHT, Warren Wright first shows up in the Westford occupational listings in 1860 as twenty-two-year-old sash-and-blind maker. Then living with JOHN HOLMES and his brother Henry in Holmes's house, Warren probably was employed then by DWIGHT CHESTER in Chester's factory. Although listed in the 1865 census as a mechanic, Warren had enlisted in the First New York Engineers in September 1864 and was then in the Union army. He must have returned briefly to the township because his daughter Jennie and his wife, Ann (niece of HEZEKIAH SKINNER and the daughter of Lucinda Skinner and Delaware County carpenter Jacob H. Wright, who was no relation), were living there then with her mother. The Wrights seem not to have returned to Westford again.

61. WILLIAM E. WRIGHT (WN: 1841 – 1925)

1850 1855 1860 1865 1870 1875 1880 1890 1900

Worcester N: Worcester—WFD—Maryland—WFD—Binghamton—Worcester—WFD—?—WFD

As the son of MILTON WRIGHT, Will Wright was related to all the other carpenters of that family in Westford, as well as to his cousin JUDSON WINANS. He also was tied to JAMES and JOHN HOLMES through his marriage to their sister Sarah. Wright first reached employment age during his father's stint as a farmer, and thus he did not emerge formally into the craft until the 1870 census listing, although the Holmes diary briefly shows him at work in 1869. Later in that year, he and Sarah and their children had moved to Schenevus, presumably so that he could prosper there in his relatively new career. The family was building a new house in that village during 1869, as the local paper reported, and in 1871 they began another one. Wright reported owning real estate worth $2,500 in the 1870 census, so that at first glance the move to this settlement on the new railroad indeed seems to have been shrewd. But in 1875, they moved back to Westford, where—shades of his father's life—Will and Sarah were farming, although Will was to claim no occupation in the 1880 census. The family remained in Westford through 1884, when they left for Binghamton, and after Binghamton they lived for a time in Worcester. Will and Sarah were again back in Westford, without their children, by 1900. Following Louisa's death in 1905, Will seems to have left again before showing up, alone at age eighty-three, in the 1925 Westford census. He died later that year, the longest-lived of the carpenters who had practiced in the township in the heart of the nineteenth century.

Appendix 3
The Holmes Memoranda

These memoranda were jotted by James C. Holmes on the undated pages at the end of his 1869 diary. They are printed here in a modernized form (see the beginning of chapter 2 regarding my editorial changes to the diary proper).

[1]

Hemlock lumber sold	to J. Winans	3,264 feet
	to J. Winans	140 feet
	to J. Winans	107 feet
Used by self	Sills	472 feet
Used by self	Joists	385 feet

[2]

Received of C[larissa] E[lizabeth] Holmes

Oct. 6	1866	100.00	Paid 14.00
Apr. 1	1868	100.00	
Sept. 1	1868	50.00	
Jan. 4	1869	50.00	

April 1870 Paid H[orace] W. Holmes 25.00

Apr. 2 H. W. Holmes
Apr. 2 1869 150.00

Lyman Snyder
May 20 1869 100.00

[3]

Received of J. Kelso

Dec. 15 1869 200.00
1870
Apr. 2 Mary A. Snyder 100.00
Sept. 11 1869 J[ames] B. S[kinner] 100.00
JBS sold [note?] to G[erman] Baldwin

[4]

Date	[Entry]	
Jan. 9	J. Winans Dr*	
9	to 30	hours on house
16	to 21	hours on house
22	to 39½	hours on house
30	to 15½	hours on house
Feb. 1	to 7½	hours on house
13	to 19	hours on house
20	to 20½	hours on house
24	to 24	hours on house

177† + 83 = 260 hours = 26 days 2.50 $65.00
83

Cr
Mar. hours
27 45‡

*Holmes used *Dr* for Debtor and *Cr* for Creditor.

†This figure derives from those listed above; the figure 83 listed immediately after it (and repeated underneath) apparently represents Holmes's hours of labor on the Winans house late in 1868. Holmes at some point also overwrote another figure (either 241 or 247) here, but he used 177 in drawing his totals to the right.

‡This credit to Winans may represent his work in the Holmes shop in March 1869, although the hours credited to him for that labor in the diary proper do not match this figure.

[5]

Date	[Entry]
Apr.	Everett Herdman Cr
17	By 5 days (24) 5½ d + 6*
May	(1) 6 (8) 6 (15) 6 (22) 6 (29) 6
June	(29) 6 (5) 6 (12) 6 (19) 5½ (26) 6
July	(3) 6 (10) 5 (17) 5½ (24) 6 (31) 6
Aug.	5 + 6 + 6 + 6 + 2
Sept.	3½ + 6 + 4 + 6 + 4
Oct.	5¾ + 6

[6]

House for Self	
Bills at Fox & Barnes	85.45
30 loads Stone	60.00
Basswood Lumber	10.00
400 Brick	6.00
Shingles	25.00
Siding and Lumber	13.00
Window and Doors	15.00
Bill at Fox & Barnes	13.51
Shingles	30.00
Paint and Oil	20.00
Siding of Gardner	17.50
Sand	15.00
40 80 bushels Lime	40.00
Tyler for stone	5.00

*These entries are simpler than they may appear. This line credits Herdman with 5 days' labor as of April 17, then adds a credit for the week ending the 24th (5½ days) and one for the week ending May 1 (6 days). On the next line, Holmes at first entered the May 1 credit before moving it back to the previous month; then he indicated for each week (the numbers in parentheses are the dates of each Saturday) the number of days worked. For August, September, and October, he dropped the dates and simply entered the number of days. These figures tally with the entries in his diary over these months.

Dressing Lumber	3.50
Bill at Barnes & Fox	47.00
E. Snyder [for] plaster	8.00
300 lbs. Nails	15.00
M. J. Hubbard	15.00
D. Salisbury	4.00
G. Baldwin	65.00
J. Goodenoughf [*sic*]	2.00
W. O. Ashley	126.57
4,000 feet hemlock	50.00
60 bunches lath	25.00
Glass and Putty	12.00
5 doors	10.00
Trim for 9 10 doors	10.00
100 Days' work	250.00
[Grand total]	[$998.53]

[7]

Ashley's Bill	Doors Windows and 73 feet blinds	105.03
Hubbard Bill		93.75
Baldwin's [job] About		
9 windows		30.00
		228.78
		.05
		11.4390*

*Holmes was here figuring either his markup on these items (ordered for these clients from Barnes & Fox, presumably) or the discount he would receive from his supplier and perhaps was planning to pass on to Ashley, Hubbard, and Baldwin.

Notes

INTRODUCTION

1. Holmes probably went to Albany in mid-December and returned to his hometown of Westford on January 2, 1869. Of course, he could have purchased the diary nearer home: on December 8, the *Schenevus Valley News*, the paper in the village where he caught trains and did a good deal of shopping, noted that the firm of Gurney and Cass there had "a large quantity of Diaries for 1869, which are worth a notice."

2. Organized in 1852, the A & S did not complete its Albany to Binghamton route until 1869. Because of its potential as a coal carrier, the completed line soon attracted the interest of Jay Gould and James Fisk, Jr., who had gained control of the Erie earlier in the decade and were locked in a difficult competition with Vanderbilt's Central. The chicanery and violence that marked their struggle to secure control of the A & S are chronicled in Charles Francis Adams and Henry Adams, *Chapters of Erie* (originally published 1886; Ithaca: Cornell University Press, 1956), pp. 137–93.

3. Several recent studies exemplifying the use of documents for the study of craft activity deserve special mention here: Catherine W. Bishir, "Jacob W. Holt: An American Builder," in Dell Upton and John Michael Vlach, eds., *Common Places: Readings in American Vernacular Architecture* (Athens: University of Georgia Press, 1986), reprinted from *Winterthur Portfolio* 16:1 (1981); Peter N. Moogk, *Building a House in New France: An Account of the Perplexities of Client and Craftsmen in Early Canada* (Toronto: McClelland and Stewart, 1977); Charles F. Hummel, *With Hammer in Hand: The Dominy Craftsmen of East Hampton, New York* (Charlottesville: University Press of Virginia for the Winterthur Museum, 1968); Edward B. Fix, "A Long Island Carpenter at Work: A Quantitative Inquiry into the Account Book of Jedidiah Williamson,"

Chronicle of the Early American Industries Association 32:4 (1979) and 33:1 (1980); and James M. Denny, "The Vernacular Building Process in Missouri: The Building Activities of Nathaniel Leonard," a paper delivered before the Twenty-fourth Missouri Conference on History, April 16, 1982. Of more general interest are Tracy Kidder, *House* (Boston: Houghton Mifflin, 1985), and Bob Syvanen, *What It's Like to Build a House: The Diary of a Builder* (Newtown, Conn.: Taunton Press, 1985).

1. THE WORLD OF JAMES HOLMES

1. Information about the life of James C. Holmes, his family, and other residents of Westford and the surrounding area in the nineteenth century has been derived from a number of printed and manuscript sources. These are listed in the "General Note on Sources."

2. J. H. French, *Gazetteer of the State of New York* (Syracuse: R. Pearsall Smith, 1860), p. 530.

3. Presumably, Holmes had produced maple sugar in 1874, too, and had also then owned the horses and the bulls; presumably, in 1875 he still owned the poultry and would have had an apple harvest of perhaps larger size. But the census report covers, except for special items, only the period from June to June.

4. *Schenevus Valley News* (10/19/1869): "Hops are worth from 18 to 20 cents, with a good demand for the article.—A good many farmers are holding for 25. Large quantities are being shipped now."

5. This description of Holmes's death is based on interviews conducted in October 1983 with his great-grandnephew, Lewis J. Holmes of Westford.

6. Westford Baptist Church, Baptist Book B, collections of the New York State Historical Association, Cooperstown.

7. These calculations use the figures supplied in *Census of the State of New York for 1865* (Albany: C. Van Benthuysen & Sons, 1867), pp. 265–66, and *Census of the State of New York for 1875* (Albany: Weed, Parsons & Co., 1877), pp. 259–60. In 1865, the county was reported to have 9,999 dwellings; in 1875, it had 11,013. The Town of Oneonta accounted for 442 of the increase; Westford rose modestly from 290 to 314. More will be said of these trends in chapter 4.

8. Actually, a statewide directory apparently canvassed for in 1869 lists Winans as practicing the craft in Westford. See *The New York State Business Directory, 1870* (Albany: Sampson, Davenport & Co., [1870]), p. 439.

9. U.S. Bureau of the Census, 10th Census, *Report on the Statistics of Wages in Manufacturing Industries*, Joseph D. Weeks, comp. (Washington, D.C.: Government Printing Office, 1886), pp. 535–43, 560–63. It needs to be stressed here that the carpenters surveyed for this report were *not* employed in

the building trades; they were working at their craft in a variety of industrial plants. Presumably, those in the building trades would have been paid at least as much.

2. THE HOLMES DIARY FOR 1869

1. John T. Hoffman (1828–1888) was elected on the Democratic ticket in the notoriously corrupt election of 1868, which set the stage for the Tweed Ring scandals of the early 1870s.

2. The *Christian Advocate and Journal and Zion's Herald* had been established in New York City in 1830.

3. Joshua Cross was supplying Holmes with shingles. Cross apparently was not a relative of Mrs. Coats and in 1870 was not listed as a resident in her house. Perhaps he was currently in her employ, or perhaps Holmes was paying off a debt Cross owed Coats by giving the money directly to her.

4. Evidently a meeting connected with the Union Library, which was presided over by George Skinner.

5. During 1869, Holmes filed perhaps as many as seventy-four saws, most of them used for cutting firewood. His income from this source may have totaled about $28, or the equivalent of eleven days' labor as carpenter. This was not a negligible supplement to his income, especially since much of the filing occurred during the cold months, when wood saws were most in need and when Holmes would not always be able to secure craftwork. Here was one part of his composite economic life.

6. The Cash Account indicates a payment of $.25 to Skinner but does not explain its purpose.

7. Holmes was buying shingles from Joshua Cross and selling them to Winans. Shingles (or shakes) were usually cut in widths ranging from 2.5 to 14 inches and were packed four bunches or bundles to the thousand—a "thousand" being the equivalent of 1,000 shingles 4 inches wide, for a total of 333.3 lineal feet. With a 4-inch exposure, this quantity would cover about 125 square feet of roof surface. See William A. Radford, *Radford's Estimating and Contracting* (Chicago: Radford Architectural Co., 1913), pp. 245–46.

8. Although at this time she was living in the Hudson valley, Clarissa Holmes was visiting in Westford.

9. Evidently Horace Greeley's New York paper, founded in 1841. Although the Democratic party was strong in Westford, Holmes apparently was a Republican.

10. The *New York Examiner and Chronicle* was a Baptist paper founded in 1824 as the *Baptist Register* and known under several other names before becoming the *Examiner and Chronicle* in 1865.

11. The Sash and Blind Factory in Colliersville was owned by Simeon R. Barnes (b. 1832) and Francis Marion Fox (b. 1829) of Schenevus. Both Barnes and Fox had been making sash and blinds in Schenevus as early as 1850, perhaps independently; by 1855 they had become partners there and were builders as well as lumber merchants. The partnership was dissolved in 1875, after which Barnes was a prominent Otsego County contractor, an officeholder, and eventually a resident of Oneonta. Fox, the son of a carpenter and builder in Cortland County, continued the business in Schenevus and Colliersville after breaking with Barnes. See *Biographical Review . . . Biographical Sketches of the Leading Citizens of Otsego County, New York* (Boston: Biographical Review Publishing Co., 1893), pp. 105–6, 696–97.

12. Holmes is referring to Madison (later Colgate) University in Hamilton, New York, a Baptist college founded in 1819. Smith, a fundraiser for that institution, had been canvassing in Schenevus the previous week. See *Minutes of the 74th Anniversary of the Otsego Baptist Association* (Cooperstown: *Freeman's Journal* Office, 1869); *Schenevus Valley News*, 1/19/1869, p. 2; *The First Half Century of Madison University (1819–1869)* (New York: Sheldon & Co., 1872), p. 408.

13. The Westford Literary Institute, founded in 1859, was a coeducational academy just at its peak in the late 1860s. Operated in a Classic Revival building formerly used (at its old location across the way on the west side of Elk Creek road) as the Methodist Episcopal church (fig. 18), it was being run in this period by members of the Draper family, including Andrew Sloan Draper, later famous as the New York State Commissioner of Education and the President of the University of Illinois. Among its students in 1866 was Arthur Holmes, James and Mary's son. The Institute was not being run in 1869.

14. *Tally* probably was lapped lumber of some kind, either for underlayment on the roof or for partitions. I have not found the term in various dictionaries or contracting handbooks, but it is verified in other manuscripts; for instance, the account book of John J. Rose of the Town of Maryland, now in the collection of the New York State Historical Association (NYSHA), has this entry under December 3, 1863 (p. 39): "Rec'd for 2000 feet of pine tally boards, @ 9/[shillings per hundred feet], $22.50."

15. In the "House for Self" summary of expenses, Holmes listed the cost of the 400 bricks probably referred to here as $6.00. A cash account entry on March 1 ("John Hubbards by S. Manzer, $6.00") suggests that these bricks were purchased from John Hubbard, a Schenevus brickmaker. They were, then, local items, not products brought in by rail.

16. This was the first of two orders of machine-made items, probably for Holmes's own house.

17. Daniel Nash Allen (b. 1807) died of consumption. He had farmed in Westford since 1842. Late in November 1868, probably because of failing

health, he had held an auction at his house on Darling Hill, south of Westford village, and had sold his 102-acre farm there to Harrison Groff (*Schenevus Valley News*, 11/24/1868, 2/16/1869).

18. This would have been the bill for the items ordered on February 1.

19. Evidently, the screws and bolts were for Kelso's wardrobe.

20. It would appear that Wright took a load of lumber to Schenevus for Holmes, perhaps to Ferry's mill.

21. John K. Tyler soon was named the Republican candidate for supervisor; William H. Platner, the Democrats' choice, won the election held later in February.

22. *Casing* refers to the trim around windows and doors; *base* is baseboard.

23. Mary Wright Griggs died on September 11, 1868, at the age of fifty-six. Predeceased by her husband, John R. Griggs (1805–1859), in her will she named her son Arthur executor of her estate and guardian of her minor children. When Holmes accompanied Arthur to the county courthouse in Cooperstown on this day, he also took the opportunity to clear up some personal business, filing three deeds in the clerk's office. The records of the Otsego Surrogate Court (lib. 15:140–43) contain Holmes's later testimony, given in Cooperstown on March 9, that he had witnessed the signing of the will on September 7, 1868.

24. No local record of such an auction has been found; it may have been related to an imminent dissolution of partnership between brothers Waldo H. and John K. Tyler. Waldo had become the partner of John T. French in a general store in Westford in 1860; French left that partnership in 1862, and John K. (Ken) Tyler entered the business sometime thereafter. Waldo bought Ken out in April 1869. On Holmes's considerable dealings with the Tylers during the current year, see note 43.

25. Holmes's skill as a finish carpenter is suggested here.

26. Myron Hubbard apparently wanted Holmes's advice on the ready-made elements that would have been purchased for his new house from Barnes & Fox. Ashley probably accompanied them for similar reasons.

27. Bilious fever was defined in George Shipman's *The Homoeopathic Family Guide* (Chicago: C. S. Halsey, 1865) as the same as remittent fever: "In this disease, which commences with a chill, the fever is continued, but once at least in the twenty-four hours, it becomes lighter or *remits* [Shipman's italics]. . . . In this disease, with the other usual symptoms of fever, there is more or less distress in the stomach; the tongue is white or yellowish, sometimes brown; vomiting is generally present; there is headache, and more or less thirst" (p. 46).

28. In April 1868, Daniel Aldrich bought from James and Mary Holmes for $2,150 a parcel of 26 acres, but he resold it to them in March 1869 for $2,000 (see 3/16 entry). The settlement with Aldrich probably pertains to this land exchange. Since the parcel in question was purchased by Holmes in March

1868 from a kinsman for only $500, it seems likely that he had arranged to buy bare ground and before actually purchasing it had built a house on it. After repurchasing the property, the Holmeses kept it for another ten years and then sold it to his brother Horace.

29. The work referred to here most likely was the wardrobe Holmes made for his uncle in January.

30. This was paid back early, on March 23.

31. No direct record of such a sanity hearing, which must have been held in Westford proper, has been found. But the *Proceedings of the Board of Supervisors of the County of Otsego for 1878* (Richfield Springs: "Mercury" Book and Job Steam Printing House, 1878) indicates that a Danverre Drake was listed as a pauper at that time in the County Home in nearby Middlefield (p. 14), and that a Damon Drake (who was from Westford) was confined in the Willard Asylum (p. 19). Neither of these individuals shows up in the 1865 or 1870 Westford population schedules, however.

32. See note 23.

33. As suggested in the Introduction, this move could not have been to the new Holmes house, no part of which was yet standing. What precisely is intended is not clear. He may have been using part of his present home, just behind the Mel Griggs store on the Clarksville road, as a work area; or perhaps the family had closed up part of that dwelling for the winter and was now moving back into it.

34. Manure accumulated in dairy barns and horse stables and in their yards over the winter. Cleaning it away and spreading it on the fields was typically a task of the early spring. Arthur Holmes was performing this on the family's old five-acre farm (or at the barn Holmes was renting from Susan Chester; see 4/13 entry), not on their new one, where the barn was not to be built until later in the year.

35. A boring machine was a double-crank (i.e., two-handed) bit drill used, among other things, for preparing the mortises in a heavy timber frame. See Paul B. Kebabian and Dudley Whitney, *American Woodworking Tools* (Boston: New York Graphic Society, 1978), fig. 43.

36. The nature of this settlement is not clear. Since no land had been exchanged between the brothers up to this time, and no mortgages ever were filed between them, the best possibilities would be that Joseph had borrowed money from James (unlikely, in view of the fact that Joseph was much better off than James) or that James had completed a job for his elder brother in 1868.

37. As mentioned in note 28, Daniel Aldrich had bought a piece of land in the north end of the village from James and Mary Holmes. Now they were repurchasing it from him.

38. It may be assumed that Winans has begun to work for Holmes at this point, although the record of his hours is very complicated and it is conceivable

that he here was working in Holmes's shop but was making items for his own new house. On the other hand, why Holmes would record the number of hours Winans worked on the latter job is unclear. See chapter 1.

39. The settlement, as the Cash Account makes clear, was for the money Holmes's brother-in-law had borrowed from Holmes on March 6.

40. It seems likely that Wright, who withdrew from the Westford Baptist church on April 3 with the intention of joining the one in Schenevus, was in the process of moving to the latter village. His borrowing of Holmes's horse on February 5 and again on March 29 and April 12 probably was related to this impending move. Perhaps his borrowing of $15 from Holmes on March 6 also was related to the move.

41. The underplayed manner in which Holmes here records two such disparate events—with nearly the same language and in adjoining lines in the diary—is a good mark of his typical practice as a diarist.

42. Another touch of the laconic: Holmes's evaluation in this last phrase may refer to the difficulty he felt in taking his mother-in-law, Marinda, and his sister-in-law Orilla to David Shelland's funeral without the company of his own wife, Mary. Or it may simply refer to the condition of the hilly road he had to pass in going direct to Worcester; roads in this season were characteristically in bad shape, and the storm that had lasted all the previous day certainly wouldn't have improved their condition. Or it may refer to both. (The spelling *Sheldon* used in the first sentence here is reported by *Biographical Review . . . Biographical Sketches of the Leading Citizens of Otsego County, New York* (Biographical Review Publishing Co., 1893), p. 809, as the Scots original of the name, which was modified in America to *Shelland*. Why Holmes used both in adjoining entries is by no means clear; nor is the way the name was pronounced within the family itself.)

43. April 1 was a quarter day, and probably the most important one in the year since it was when renters of village dwellings or of farms traditionally moved. The items noted here represent Holmes's attempt to settle long-outstanding accounts. His dealings with the firm of the Tyler Brothers present the most interesting case. The second item with regard to them marks a simple settling of Holmes's running account with their store. But the first is quite different. This is the second of a series of ten payments, totaling $243.50, which Holmes received from the Tylers (or, following the dissolution of the firm in April, from W. H. Tyler) between January 15 and July 16. Since it is equivalent to almost one hundred days' full wages for Holmes as carpenter, this total is quite large. What it represents is not clear. It is possible that Holmes completed some major job for the Tylers in 1868 and was being paid piecemeal for the work. But it is also possible that they were financing his own new house, giving him cash seriatim to cover his costs.

44. These two payments require some comment. Holmes had bought 5½

acres of his new farm from Phoebe D. (Dolly) Bentley on April 28, 1868 (Otsego Co. Deeds, lib. 149, p. 514), for $400. His payment of $28 would represent a year's interest on this amount at 7 percent, a rate commonly in use in the period. Robert S. Hall, the Westford miller, produced lumber as well as grains; the mill's previous owner, George Guy, reported in the 1860 census that he had turned out 120,000 board feet, probably part custom work and part for direct sale. It seems clear from this entry in the diary that Holmes was having Hall do custom sawing of timber or rough lumber, almost certainly hemlock (memoranda 1 and 6), which he had purchased locally for his own house.

45. Holmes probably was moving these bulky items from his old house to his new one.

46. Judson Winans repaid this loan on April 8.

47. Later annotations entered by Holmes in his 1869 diary indicate that in April 1870 he paid his brother Horace only $25 on this debt.

48. Church meetings were usually held on the first Saturday of the month. In the clerk's record in Baptist Book B under this date, Stephen J. Cottrell made the following entry (NYSHA Colls., p. 125):

> Regular Church meeting to day, and a large number in attendence, all exhibiting a very lively spiritual interest. Some Brothers and Sisters coming out to the meeting that have not meet [*sic*] with us before in some time. After the conference the following business was transacted:
>
> Bro. O. J. Prine was elected Dea. in accordence to a resolution of the previous Church meeting [this had simply called for such an election at the next "covenant meeting" of the congregation].
>
> Dea. P[omeroy] Wright and Br. F. B. Gardner were appointed to visit Bro. James Badeau and report to the Church meeting [this visit was a disciplinary effort].
>
> Dea. Flavel Wright, Bro. Joseph Esmay, and Bro. James C. Holmes were appointed Pulpit Committee for the ensuing Year.
>
> Bro. Wm. E. Wright requested a letter of recommendation and dismissal for the purpose of uniting with the Schenevus Church.
>
> Voted that his request be granted.

49. Perhaps Holmes had heard rumors that Hubbard Fox was to be reassigned. In fact, he was to stay in the Westford church until the following spring. It should be noted that at this time the Baptist church in Westford had no regular minister. This fact explains the need for the pulpit committee, to which Holmes was appointed on the previous day: it was the committee's job to secure visitors—or a permanent minister if one could be found. It also explains Holmes's attendance at the various other churches in the town in 1869.

50. Holmes apparently made this ladder for his own use rather than for sale.

51. That is, Holmes used Tyler as his agent in making this payment on the millwork he had ordered for his new house.

52. *Blinds* were shutters. Twelve sash normally would make six complete windows, although some windows on the second floor of his new house were single fixed sash rather than double hung ones. Where he got these sash isn't clear. Since he first notes picking up sash in Schenevus on April 22, he already may have had on hand the ones painted on April 10, and they thus may not have been part of his Barnes & Fox orders of February 1 and February 5. But perhaps these sash, like those "received" on May 7, were delivered to him rather than picked up; if so, they might well have been part of those orders.

53. Near Susan Chester's house on the northwest side of the village's main street was a large barn. Holmes apparently was renting it as a temporary storage place until his own new stable and barn were completed.

54. Holmes is fairly careful in his use of the terms *wood* and *timber*. By the former he seems almost always to be referring to firewood; the timbers he hewed on this day probably were milled framing members that he was readying for the frame of his new house by drilling and chiseling mortise-and-tenon joints in them.

55. Holmes kept close count on Herdman's time; see memorandum 5. Although hired by the month, in fact Herdman was expected to work a set number of days each month (a month was composed of four six-day weeks). The details of the economic dealings between the two men are given in chapter 1.

56. The Cash Account entry for this date indicates that Holmes paid his sister Sarah $1 for some unspecified item purchased for his wife, Mary.

57. Denton Salisbury, who lived next to Holmes's new house, apparently was collector of school taxes during 1869. See also note 89.

58. Holmes is here speaking of the new shop that he was preparing to erect and enclose across the road from his new house. See chapter 3.

59. It isn't entirely clear who Ashley's "man" or "hand" was. In the 1867 assessment book for the township, an Abirham Man, Jr., was listed as a landowner, although this identification seems rather unlikely. Holmes uses the name John on May 12, but there are simply too many young laborers of that given name listed in the 1870 census for any plausible identification to be made on this basis alone. Although John Goodenough is possibly the man in question, that Holmes named him specifically on May 6, and on June 2 noted paying him for his labor on that one day a month earlier, suggests that in fact the unnamed man involved in the present instance was someone else.

60. Holmes may have been referring to young Alden Chester, Jr. (WN, 1848–1934), the son of the deceased cabinetmaker and his widow, Susan Draper Chester, and later a justice of the New York State Supreme Court. Although young Chester had taught in the Westford Institute in 1868 and was soon to return to the area, where he was to serve for a time as the editor of the *Schenevus Valley News*, he was at this time in Boston (David Addison Harsha, *Noted Living Albanians* [Albany: Weed, Parsons, & Co., 1891], pp. 54–61; *New York Times*, 2/13/1934, p. 20). It thus seems likely that Holmes in fact was

referring to Susan Chester by her husband's initial (the reference, that is, is to "Mrs. Coats and [Mrs.] A. Chester").

61. Milton Wright was just at the point of moving to Schenevus, whither he was following his son William and William's wife, Sarah Holmes. There both father and son soon were to be active in the building trades. See chapter 5.

62. This perhaps was part of a settlement that the heirs of Calvin Holmes had made with Anna Howe Holmes prior to her departure from the town. Calvin's children shared equally in the family farm, although Horace Holmes and his wife, Lucinda Skinner, in a deed dated 1/7/1870 but not filed until 12/14/1876 (Otsego County Deeds, lib. 175, p. 537), bought out his siblings for $2,000.

63. This may have been a loan from Snyder or could have been payment for work Holmes did for him at some previous time.

64. A *backhouse* was either the outhouse or the rear shed attached to a house. That Holmes built this one in a single day makes it clear that it was an outhouse.

65. Holmes indicated in his cash account for May 18 that he had received these amounts from the two men; apparently they were short-term loans, which the money received from Lyman Snyder on May 20 allowed him to repay.

66. William O. Ashley, according to the minutes of Oyer and Terminer (5/31/1869, 2: 268), had been called to serve on the Otsego County grand jury. While at the courthouse, he recorded two deeds for Holmes; the $3.80 was for this purpose.

67. This was a first payment on the pigs purchased on May 27.

68. Farmer William O. Wilbur owned land adjoining the parcel Holmes had repurchased from Aldrich in March. It was on the common line between these properties that Holmes and Wilbur were dividing the responsibilities for fencing.

69. The Worcester Baptist Association, of which the Westford congregation was a member, was holding its annual meeting in Clarksville at this time. According to Baptist Book B, however, Judson Winans was not one of the official Westford delegates.

70. The ironic tone here should not conceal the fact that as a member of the present pulpit committee, Holmes himself was partly responsible for the stranger's assault on religion.

71. The *Schenevus Valley News* quoted the *Cooperstown Freeman's Journal* on July 6 as saying that "the recent warm rains" in the area had not been good for the hop growers. Its declaration on July 13 that "beautiful growing weather" had returned was, as Holmes makes clear, premature.

72. Draper's Hall was the large Classic Revival structure in which the Westford Literary Institute had been housed. It is illustrated (in its present rather deteriorated state) in Diantha Dow Schull, *Landmarks of Otsego County* (Syra-

cuse: Syracuse University Press, 1980), fig. 219, as well as (in a contemporary photo) in figure 18. Schull's note regarding the name of the structure would seem to be in error; if it was being called Draper's Hall in 1869, as Holmes indicates, it hardly could have been "renamed after Andrew S. Draper" following his fame as New York State Commissioner of Education and President of the University of Illinois at the end of the century. Andrew S. Draper, who had left Westford with his family as a boy but did return to teach briefly in the Institute, was the nephew of its founder, Milton Draper, and Milton's siblings Fernando P. Draper and Susan Draper Chester (see appendixes 1 and 2). For a considerable period of time he was a law partner in Albany with his cousin Alden Chester (note 60). See Harlan Hoyt Horner, *The Life and Work of Andrew Sloan Draper* ([Champaign]: University of Illinois Press, 1934).

73. Bentley Hollow was an area north of the village where many people of this name were settled. It was the site of a Methodist Episcopal church.

74. Jacob Tipple had just recently purchased a house near Holmes's new house. Holmes evidently was renting Tipple's barn for storage of the hay that he was about to purchase from George Wing (7/21 entry) as well as that which was to be cut (by Daniel Aldrich, Everett Herdman, Judson Winans, and perhaps Holmes himself) on his own new farm near Tipple's place over the next two weeks.

75. A Westford correspondent wrote in the *Schenevus Valley News* of August 3 that the Methodist Episcopal church there had hung a "toned" bell costing $325 in its belfry the previous week.

76. Holmes left the blank here. He may have been referring to the Welsh native H. O. Rowland, whom the Westford Baptists had induced to preach for them during his vacation from Madison University.

77. The Cash Account notes a payment in this amount to "Picture Man." Holmes may have been arranging for photographs to be taken of his family, but it seems more likely that he was purchasing prints of some sort for his new house.

78. Holmes actually wrote $30.93 here, evidently a mistake; in his cash account, where he recorded selling the butter to produce dealer Fernando P. Draper, he wrote $31.90. This amount, substituted for his error in the text, is probably the right one; it would work out to be $.30 per lb. In any event, one should note how high this price was in a period when Holmes's daily wage was a mere $2.50. It also should be noted that we have here likely evidence of Mary Holmes's contribution to the family's economy. Although decisive information about how he used his own income from the farm is not available, it may be that in this case, as in those studied by Deborah Fink in twentieth-century Iowa, the farmwoman's income provided for the family's daily expenses, whereas her husband's profits were consumed by the costs of establishing and running the farm. Holmes's income from carpentry, of course, complicates the

issue here. See Deborah Fink, *Open Country, Iowa: Rural Women, Tradition, and Change* (Albany: State University of New York Press, 1986), p. 56.

79. Holmes appears to have been joking here. On April 24, the *Schenevus Valley News* reported, "Summer has just come. The warmest days of the season were Thursday and Friday last."

80. The firkins would be used for packing butter for sale.

81. The Otsego County Fair, which was one of the first of the modern agricultural fairs established in the United States, dated from 1817. It was running this year from September 14 to September 16. The draft Holmes bought was for $67 and was debited against D. Chester in the Cash Account. Apparently on this occasion and on December 24, when he sent Dwight Chester $50 and soon received Chester's note in return, Holmes was investing the money or banking it with Chester. Although the note Chester sent Holmes might have been a canceled record of the latter's indebtedness to him, Holmes's language in the 12/27 entry suggests the former interpretation.

82. The mortality schedule in the 1870 Westford census does not list Mason's death; he may have died elsewhere, although Holmes's language clearly suggests that he learned of the death the day it occurred.

83. This probably was Rowland's last sermon in Westford during his vacation (he visited again at Thanksgiving, when he already may have begun preaching in Middlefield). Apparently the congregation, perennially unable to keep a minister, liked him and took his impending departure hard. Church clerk Stephen J. Cottrell noted in Baptist Book B the next Saturday that there was poor attendance at the stated meeting held then because "all [were] feeling the loss of our beloved Brother H. O. Rowland, who returned to Madison University to persue [*sic*] his studies."

84. The Schenevus Valley Agricultural Society, organized in 1865, was holding its fifth fair on its grounds at the west end of the village.

85. John K. Tyler had started an insurance business after splitting up the partnership with his brother Waldo in April.

86. Mary Jane Pank (b. 1842), the mortality schedule for the 1870 census reports, died of consumption. But the presence of an eleven-year-old grandchild in the household of her husband's parents (Robert and Mary Pank) in the 1880 census suggests that childbirth may have been a contributing, if not the major, factor.

87. Two comments on Holmes's accounting might be of use here. His "1 day [plus] 2 hours" in the first line indicates nicely how exact a unit of measurement the "day" was for him; it was, in fact, ten hours, so that on this occasion he worked twelve hours. The second point is that his settlements with Winans and others must be presumed to have rather precise terms. Sometimes, however, those terms are not completely indicated in the record. Here, for instance, "settled up to Baldwin's job" might be read so as to include that job or to in-

clude everything prior to that job but not that job itself. These questions are dealt with in my text proper.

88. This water wheel may have been the power source for the cider mill that OTCO.72 reported the Markses to be running. Their choice of Holmes to work on it, although perhaps explained in part by kinship ties, suggests in yet another way the range of his skills as a carpenter.

89. According to state law (see Victor M. Rice, *Laws of New York Relating to Common Schools* [Albany: Weed, Parsons, & Co., 1868]), each school district had to elect at least one trustee to oversee its operations. Many districts still tended to be rather small—the consolidation of districts into graded schools was just beginning—but Holmes was elected trustee for the village school (which served the third district), so he was responsible for some ninety-nine students. His grandfather, John Kelso, had been one of the initial trustees of the same district, and over the years several members of his family had been very active in running it; indeed, his father, Calvin, had first come to Westford seeking work as a teacher in the school, which Calvin's brother John P. Holmes was then serving as a trustee. (See the manuscript "Records of School District No. 3, Westford, 1813–1836," NYSHA.) James Holmes's report, filed December 13, 1870, gives specifics on his duties in 1869–1870.

90. As I remarked in note 57, it seems that Denton Salisbury was serving as district or town school tax collector in 1869. It was part of the trustee's duty to make out the tax list for a given district, which was then given to the collector for action. The 1868–1869 trustee for District 3 was the carpenter Erasmus Snyder.

91. It is unclear just how Holmes thought he had "lost" those hours. Herdman had worked a total of more than 150 days for Holmes since he was hired; 2½ hours would be a very small fraction of that total. Herdman's compensation is discussed in chapter 1.

92. This was a loan from Griggs; see 11/12 entry.

93. This cow was part of the settlement with Herdman; see chapter 1.

94. Although a nursery was being run in Schenevus by William W. Cone, Cone presently was visiting Kansas, about which he was writing letters to the *Schenevus Valley News*. Perhaps because of his absence, Holmes and Winans bought their apple trees from Edwin C. Frost (b. ca. 1808), a horticulturalist and tree agent from the village of Watkins in Schuyler County (see Hamilton Child, *Gazetteer and Business Directory of Chemung and Schuyler Counties for 1868–1869* [Syracuse: Journal Office, 1868], p. 235). Frost apparently was selling nursery stock in Schenevus late in 1869, probably as an itinerant or through some local agent.

95. Bounties were authorized for enlistees during the Civil War. After the war ended, draftees who had paid commutation fees or had hired substitutes were in some instances awarded similar payments. Such a bounty had been

voted for fifteen Westford men at a previous meeting that the town had claimed, in court, was illegal. The town lost its appeal the past summer and now was arranging to meet its obligations (*Schenevus Valley News*, 7/6/1869). On the general issue of the bounty system, with special reference to the New York experience, see Eugene C. Murdock, *Patriotism Limited, 1862–1865: The Civil War Draft and the Bounty System* (Kent, Ohio: Kent State University Press, 1967).

Although James Holmes seems to have been absent from Westford for the period from March 1864 to March 1865, when Albert Martin was chosen acting clerk of the Baptist church in his stead, no service record documenting his presence in the army has been located. It thus seems likely that he had been drafted and either had paid the commutation fee or had hired a substitute.

96. This reading is probable, although *closet* might be read as *clock*.

97. Andrew Snyder's second wife, Emmeline R. Allen, was the stepdaughter of Fannie A. Allen—hence the payment recorded here.

98. See 10/19 entry and note 92.

99. The 1870 census indicates that Marinda Shelland was then living with the Holmeses in Westford. It seems likely that this was the time—just as winter was nearing—when she moved in.

100. This apparently was payment for the butter delivered on 11/16.

101. Daniel A. Cummings, now past seventy, would almost certainly not have been helping Holmes cut and bring in firewood. Albert D. Cummings is thus a more likely identification here.

102. "Cold, raw weather now, and winter fast approaching," reported the *Schenevus Valley News* the previous day.

103. Harman Rose (b. 1792) was not a member of the Westford Baptist church, but he and his son Edward must have attended it: why else would the Baptist preacher Ingraham Powers have come from Worcester for the funeral? Edward Rose's wife, Martha A. Rose, *was* a member of the church. She had joined in March of 1869 by a letter from the Rensselaerville congregation.

104. The child was Horace C. (Hallie) Wright, not yet a year old.

105. Holmes is here using the old shilling notation; two shillings in common practice were worth twenty-five cents.

106. Why Holmes so refers to his own sister is a puzzle.

107. In his capacity as trustee.

108. Evidently this was partial payment of the bounty claim voted Holmes in November. It is conceivable, however, that the exchange involved school monies, which by law were channeled through the supervisor in whose district the school was located.

109. The nature of this payment is not clear; as pointed out in note 81, it might represent either an investment made through Chester or a payment on monies previously borrowed from him, although the entry in the diary on

12/27 ("Recd D Chester Note $50.") does sound as if the former possibility is more plausible. In any event, the money was forwarded through W. H. Tyler, as the Cash Account makes clear.

110. That is, Waterman settled with Holmes and owed him $6; he paid Holmes off by giving him the oil and the cash.

111. Tipple (b. 1783), long a farmer in Westford but not a resident there for many years, had died on December 27 and was buried in Worcester.

3. THE MAJOR JOBS OF 1869

1. The Elk Creek valley runs roughly from the southwest to the northeast. To simplify my descriptions here, I have squared off the axis; Holmes's new house site thus actually lay southwest of the road.

2. Interview with John Silvernail, Wells Bridge, New York, October 1983.

3. This must have been the quarry still marked on the U.S. Geological Survey map of Westford, located in the vicinity of Calvin Holmes's house about one-fourth mile west of the intersection of Elk Creek road and the road to Skinnerville. The entry in the diary may be taken to mean that Holmes went with Mary and Clara to Calvin's house and then "went over to the quarry," just the wording he would have been likely to use for the quarry shown on the USGS map.

4. CARPENTRY IN OTSEGO COUNTY IN THE NINETEENTH CENTURY

1. David M. Ellis, et al., *A History of New York State* (Ithaca: Cornell University Press, 1967), pp. 173–74.

2. A convenient summary of the changes in population in New York between 1825 and 1925 is found in State of New York, *Supplementary Report on the Enumeration-tabulation of the Inhabitants of the State as of June 1, 1925* (Albany: 1927), especially the table on page 14 and the map on page 9. On the general economic and demographic decline of the region, see James Arthur Frost's excellent local history, *Life on the Upper Susquehanna, 1783–1860* (New York: Columbia University–King's Crown Press, 1951).

3. State of New York, *Census of the State of New York for 1875*, C. W. Seaton, comp. (Albany: 1877), maps g, f, h.

4. *Schenevus Valley News*, 8/3/1869. What precisely the limits of the village are in this brief story is, of course, problematic.

5. This is without counting Winans's house, built well north of the village in 1868, long before the August 1869 newspaper notice appeared.

6. *Instructions for Taking the Census of the state of New-York in the Year 1855; Issued by the Secretary of State* (Albany: 1855), p. 15. Similar stipulations appear in the instructions for 1865 and 1875.

7. The growth of Schenevus and other railroad villages is discussed in some detail later in this chapter.

8. The township of Roseboom is not included in any of these calculations because of an evident error in the 1865 figure there.

9. The number of buildings destroyed in this period is not known.

10. Because the report for 1868 is not as useful as that given for 1867, I will not use it here except in the cases of Unadilla and Worcester. The reports appeared in the *Freeman's Journal*, 1/10/1868 and 1/15/1869.

11. Actually, he made this claim in the case of several villages, with special emphasis on Unadilla's nearby rival, Otego.

12. It is difficult to estimate village population levels. The sources cited here are Edwin Williams, *The New-York Annual Register for the Year of Our Lord 1840* (New York: Stationer's Hall, 1840), and J. H. French, *Gazetteer of the State of New York* (Syracuse: R. Pearsall Smith, 1860).

13. The paper in Schenevus in the August following the Cooperstown story said that development in Otego was "brisk" and that the village was "growing finely" (*Schenevus Valley News*, 8/25/1868). As if it saw Otego as both rival and role model, the paper said similar things on several other occasions: "Otego has built up considerably this season, and is one of the finest villages in the Susquehanna Valley" (11/24/1868); "Otego is bound not to be behind her sister towns"—and hence was planning to seek incorporation (12/15/1868); "Otego has three Christmas Trees—Schenevus had none" (12/29/1868); and "The growth of this village is unabated and seems to be healthy and substantial" (1/12/1869). Such continuing references suggest that Otego had a local reputation in the period as a progressive village.

14. Although the *Schenevus Valley News* did see some marks of briskness in Unadilla (business there was "very lively," it wrote, 8/31/1869), its tone in the following passage is noticeably different from that found in its discussion of Otego: "Unadilla is not dead, neither is the odor bad from this beautiful village. Business is apparently good, and on a gradual increase" (6/22/1869).

15. D. Hamilton Hurd, *History of Otsego County, New York* (Philadelphia: Everts & Fariss, 1878), pp. 339, 240.

16. The 1840 figures are from Williams, *Annual Register*; those for 1872, from Hamilton Child, comp., *Gazetteer and Business Directory of Otsego County, N.Y. for 1872–3* (Syracuse: Journal Office, 1872), hereafter OTCO.72.

17. But Worcester no longer held the advantage. OTCO.72 lists the population of the two villages as 323 (Wor) and 393 (EWor) and the number of dwellings as 71 (Wor) and 85 (EWor). It should be noted, though, that the Worcester report for 1867 must have understated conditions there: in the case of East

Worcester, for instance, $4,000 in repairs were included, whereas for Worcester only new construction was tallied (the reporter then added that "a number of dwelling houses" also had been repaired and painted, though no dollar value for this work was given).

18. In 1840 (Williams, *Annual Register*), the estimate of Worcester's population had been 170, less than half that for Unadilla (365). In OTCO.72, the differential was even greater: 323 to "about 1,200."

19. Andrew Boyd, *Business Directory and Gazetteer of the Towns and Villages on the Albany & Susquehanna Railroad . . . 1868–9* (Albany: 1868), p. 203.

20. The cheese factory was a special mark of the railroad's influence. The operation closed, however, in 1869, and the *Schenevus Valley News* noted in that year that there was talk of converting the building to a starch factory, which would exploit the local potato crop (5/25/1869). Westford saw the opening of a cheese factory in 1868, too.

21. A steam printing press and steam power for the sash-and-blind factory both came into use in the late 1860s, marking the coming unimportance of waterpower.

22. Hurd, *History*, p. 178.

23. The 1860 figure is from French, *Gazetteer*; that of 1870, from a separately distinguished count in the U.S. Census manuscript population schedule that year; 1872, from OTCO.72; 1876, from Ashley Hotchkin, *A Concise History of the Town of Maryland* (Schenevus: 1876); 1878, Hurd, *History*; 1890, 11th U.S. Census reports, vol. 11, pt. 1, *Statistics of Population*. No separate count was made of Schenevus in 1880.

24. But the *Freeman's Journal*, 10/8/1869, had much praise for the 1869 fair in Schenevus.

25. There were several other local fair societies in the region: Unadilla had one that dated from 1857 and had been reorganized in 1866; Edmeston and Burlington townships had one founded in 1866; Otego established one in 1869. Hurd, *History*, passim.

26. The *Schenevus Valley News*, according to Hurd (*History*, p. 35), was founded as a Republican paper to oppose the *Monitor*'s turn toward the Democrats during Grant's first campaign in the fall of 1868. Oatman had been Multer's partner in running the *Monitor* (Boyd, *1868–9 A & S Directory*, p. 129).

27. As soon noted in my text, Chase was also much expanding his hardware business this year.

28. *Schenevus Valley News*, 8/25/1868.

29. *Schenevus Valley News*, 3/23/1869.

30. The "thoroughfares" in question were the Cherry Valley and Catskill turnpikes.

31. *Schenevus Monitor*, 9/27/1864.

32. The writer of the letter was Ashley Hotchkin, later author of the town

history cited in note 23 above. A somewhat less puffy response to the coming of the railroad was the "Orijunul Owed Kompoughized onto the ocashun ov furst seein the iurn Hors in the suskehanner valley, and hearin im scream: onn the advent ov his furst appeerance into ohneonta, onn the august 20 forth 1800 & sixty five. ritten kumpoughed & aranged when a standin within 2 foot ov im, an ocashunly tuchin ov im evry nough & then." This appeared in the 8/12/1870 issue of the *Pincher*, a humorous Schenevus newspaper of short duration whose "Editor and Proprietor" was Jacob Multer's son Marcus. The "owed" began

> o thou everlastin monster ov the depe!
> thou grate leviathan ov the sandy desert ov sary!
> thou behemoth ov the mountains!
> & hippopotamus ov the vallys!
> all hail 2 the! i sa thou mity beste,
> where ju come from?
> ju walk all the wa, or ju kum a foot?

How the author, "Excelsior," managed to get out of the word *hippopotamus* without a single misspelling is beyond knowing.

33. The basic information in this discussion comes from the *Schenevus Valley News*, but the particular date is given in my notes only when a passage has been quoted. Otherwise, dates are indicated in the text itself. The identification of occupations and other incidental information have been taken from the various local sources cited elsewhere in this study.

34. Since no further word of this project appeared in the paper, it apparently fell through, although the county atlas published the *previous* year, apparently taking hopes as facts, showed a house belonging to Oatman already standing on the vacant lot in question.

35. Appropriately, since Milton Wright and his son William and their families were among the founding members of the new congregation, Milton's brother Lyman, a minister currently living in Newburgh, was to be the guest preacher. Although no positive evidence of his involvement in the project has been found, it seems quite possible that Milton Wright worked on erecting the new church building.

36. Oscar S. Lane (b. ca. 1843) and William S. Hotchkin (b. ca. 1844)—husbands of Helen and Celida Wright (Milton's daughters)—were running an active business in Schenevus. Advertisements in the *Schenevus Valley News* indicate that they manufactured not only sash, blinds, and doors, but also storefronts, moldings, brackets, newel posts, balusters, and pickets. They also sold seasoned lumber and did custom sawing, scroll sawing, turning, planing, and matching and surfacing (9/1/1868). They added a 24-HP steam engine in July 1869 (8/3/1869) and were described in 1870 as careful workmen who bought all their lumber in Albany (a very important lumber market at this time) rather

than using unseasoned country products. In order to assure tight fits in their products, they had a drying room where items were further seasoned (5/28/1870). In the 1870 "Products of Industry" section of the Maryland U.S. Census, the partners indicated that they produced 1,200 doors (worth $3,600), 8,000 feet of blinds ($1,400), 9,000 lights of sash ($1,080), plus brackets and other work ($1,500), for a total of $7,580. They employed six men above the age of sixteen and one male below that age, had an investment of $6,000 in their plant and equipment, paid out $2,500 in wages, and used $2,600 in pine lumber and $500 in glass and sundries. A very bad flood that occurred July 29, 1873, severely damaged their plant, which was destroyed by fire August 21, 1875 (Hurd, *History*, p. 179; Hotchkin, *Concise History*, p. 42).

37. Mrs. Jones apparently had not been gone very long; she is listed as a Schenevus milliner in Boyd's *1868–9 A & S Directory*.

38. Apparently the shortage was more broadly felt in the county. The *Freeman's Journal* ran a story on the growing scarcity of rental housing in Cooperstown and urged its readers to become homeowners whenever possible (1/22/1869).

39. The letter from the Worcester resident appeared in the *Schenevus Monitor* on 11/1/1864; the *Schenevus Valley News* comments appeared 7/27/1869. In his *History* (p. 35), Hurd attributes the failure of the *Schenevus Valley News* to a split in the Otsego County Republican party. The paper's last owner was J. E. Young.

5. THE CARPENTERS OF WESTFORD

1. Wright's involvement in this job is evident from a mechanic's lien he filed with the Otsego County clerk on January 17, 1870, charging that Silas Rider of the village of Maryland owed him $87.79 for labor performed, according to agreement, within the past thirty days. Rider was listed in the 1870 census as a thirty-year-old farmer owning $2,000 worth of real estate.

2. Attempts to trace Westford residents listed as engaged in manufacturing in the 1820 federal census or in "manufacturing and trades" in the 1840 federal census (no such categories were used in 1830) have not suggested that the numbers of carpenters at work in the township in those earlier periods exceeded the number reported in 1855. This topic wil be discussed at greater length below. It also might be noted that no consistent pattern can be found in other Otsego townships with regard to a marked increase in the number of carpenters and cabinetmakers in 1855. It was not uniformly a boom year: Maryland went up slightly over 1850, and so did Middlefield (though neither showed the kind of increase Westford did), but only Decatur gained significantly, while Oneonta gained only one carpenter, Worcester held steady, and Otego actually lost one.

Westford's change, in other words, hardly was part of a countywide surge.

3. John R. Commons et al., *History of Labour in the United States*, vol. 1 (New York: Macmillan, 1918), p. 614, as quoted in Walter Galenson, *The United Brotherhood of Carpenters: The First Hundred Years* (Cambridge: Harvard University Press, 1983), p. 11.

4. These comments derive from information in Hurd's *History*, p. 355, from various notes in Baptist Book B, and from Lewis J. Holmes, *A Half-Century in Westford*, pp. 73–104.

5. As will be seen later, on the other hand, there *were* definite kinship/craft circles traceable in Westford even in 1855.

6. It should be remembered, however, that many of the transient carpenters practiced elsewhere for periods before and after their presence in Westford. A full record of their careers doubtless would result in considerably higher averages, although a rough count for those who have been traced before and after their Westford periods suggests that the total active career for many of them, including their time in Westford, was only around eight or nine years.

7. W. J. Rorabaugh, *The Craft Apprentice: From Franklin to the Machine Age in America* (New York: Oxford University Press, 1986), pp. 9–10; Theodore Rosengarten, *All God's Dangers: The Life of Nate Shaw* (New York: Alfred A. Knopf, 1974); for an illuminating New York example from Holmes's period, see Henry Conklin, *Through "Poverty's Vale": A Hardscrabble Boyhood in Upstate New York, 1832–1862*, Wendell Tripp, ed. (Syracuse: Syracuse University Press, 1974). The 1900 census helps one distinguish between *occupation* and *employment*—this is the crux of the problem here—by inquiring how many months over the past year each worker was not employed at his or her named occupation. The three carpenters listed in Westford that year reported being unemployed a total of fifteen months in that period (the actual figures were three, four, and eight months). William O. Ashley, who had worked on Holmes's house in 1869 and in turn had had Holmes build a new house for him, was an eighty-four-year-old mason in 1900, but had been unemployed for all twelve of the past months; his son, sixty-year-old Carlos, also a mason, had been unemployed for eight of the past twelve months.

8. The story of the Conklin family's wanderings in *Through "Poverty's Vale"* offers an instructive analog: between 1820 and 1830, the Conklins moved some ten times; from 1830 to 1834, Henry's father "worked out some and hired land to work and worked some on shares and he moved about the place from house to house every year, as mother said[,] he moved every six months" (pp. 14–15); and from this period of wandering to 1840, the Conklins "settled down"—relatively speaking, since the family still lived in three or perhaps four other houses. Yet a reliance on the census would show the Conklins as having moved only *twice* across this twenty-year period.

9. The three children, aged between four and fourteen in 1855, at that time

are all said to have been born in Broome County. But in 1865, Ann (twenty-four) and Amelia (twenty-two) are said to have been born in Otsego County, while Adline (fourteen) is said to be a native of Delaware County. Attempts to locate Jacob H. Wright in the 1840 census in New York have not been successful. The fifty-three-year-old Betsey B. Wright, in whose house Lucinda and her children were living in 1865, is said to be a native of Delaware County. She would seem to have been Jacob's sister.

10. The kinship ties extended, too, in the direction of James Holmes, whose brother Horace was to marry Hezekiah's niece Lucinda, and whose sister Fanny already had married Hezekiah's brother James B. Skinner.

11. The census-taker gave William no occupation this time, but even at twelve in 1840 he was one of the three males in the household engaged in agriculture.

12. The 1855 census says Michigan: this may be especially significant, since for New York natives the state census enumerator was instructed to give the county name, not just the state. To the federal enumerator Winans could say New York, a convenient if not an accurate answer, but he couldn't as easily specify a convenient home county. The 1875 state census, though, gives his birthplace as Livingston County, New York. His brother George, seven years his junior, was listed as born in Michigan in 1860 but was listed as born in Gennessee [*sic*] County in 1865. There are, however, counties of this name in both Michigan and New York. Attempts to find the appropriate Winans family in Michigan in 1850 proved unsuccessful.

13. Russell Winans left the church with a letter of dismission on February 21, 1836; a March 31, 1836, entry in the day book of Westford doctor Elhanan Jackson (in NYSHA) indicates that Deacon Winans held a vendue that day, surely in preparation for moving. William and Louisa Winans, Sr., and William, Jr., were granted letters of dismission from the church on April 2, 1836, and that October it was reported that they had joined the Baptist church at Lakeville, probably the hamlet of that name located at the foot of Conesus Lake in Livingston County, New York. Perhaps, then, this is where Judson Winans was born, conceivably the son of William Winans, Jr. Livingston County adjoins Genesee County.

14. Baptist Book B (NYSHA), 6/2/1855. It is clear from context that Milton Wright added this entry himself at a later time: it is unsigned but the handwriting matches.

15. Wright signed the clerk's book on both December 2, 1854, and January 8, 1855.

16. The baptisms occurred on April 5, 1857 (Baptist Book B).

17. The best example of such handiness is Nate Shaw; see Rosengarten, *All God's Dangers*.

18. The ten-year-old sister was listed as Marion (with Judson) and as Mary

(with George Winans, living in Milton and Louisa Wright's house); since this is Mary's only appearance in the Westford census, it seems more likely that there was only one girl and that she was listed twice rather than that there were twins.

19. Half of the parcel had been employed as pasture in 1864, but in the spring of 1865 Milton Wright reported owning no livestock. He was then plowing only one acre (half for corn, half for potatoes), and the remaining nine acres were wholly unimproved.

20. It was August 6, 1870, when he received a letter of dismission from the Westford church, but this was obviously received—as was often the case with such documents—after his actual departure, since the census in Oneonta had been taken in June.

21. As I remarked earlier with regard to mason William O. Ashley, workers tended to keep their occupational titles even after they ceased actual employment in the occupation in question. It also seems to be true that an emergent worker in a given occupation may well not have appeared under his proper title until some time after he in fact had begun practicing.

22. Women are not included here because their occupations are not included in the 1850 census counts, and thus the data would be incomplete.

23. Farming was the most numerous male occupation (in 1870, "keeping house" was the absolutely most numerous declaration), but the next most frequently named male occupation was carpentry. In the United States in 1850, there were 197,343 carpenters, or 1 for every 101 people. In 1890, there were 611,482, or 1 for every 102 people. In New York in those years, the totals and ratios were as follows: in 1850, 37,670 (1:82); in 1890, 73,022 (1:82). There thus was a surplus of carpenters in New York State, as measured against the national averages. To put this surplus in a different context, it might be noted that a frontier state such as Iowa had ratios of 1:124 and 1:118 in 1850 and 1890 (the carpenter counts were 1,953 and 16,201, respectively). Iowa was a better place to be a carpenter in those years, especially since the total increases in reported dwellings in New York and Iowa over the four intervening decades were quite close. New York showed an increase of 422,000 dwellings, and Iowa, 346,000. When these figures are compared to the total number of carpenters counted in each decennial census (254,000 in New York; 48,000 in Iowa), we find that the number of houses erected per carpenter in the four decades was 1.66 in New York but 7.2 in Iowa. The national ratio for the same years was 4.6 houses/carpenter. Of course, New York, considerably more urbanized and industrialized, offered construction jobs on projects far beyond the scope of those available in Iowa; it also had jobs for carpenters in industry that must have far outrun similar openings in Iowa. But as a measure of the average "house" carpenter's chances of employment at his craft, these statistics suggest

that the rural eastern craftsman was at a decided disadvantage. (Figures derived from various published volumes of the federal census, 1850–1890.)

24. *Nominal* means officially named in the relevant census.

25. Jacob Galer and William Wales are also included in table 5 because although never directly named as carpenters in Westford they had strong associations with the craft elsewhere. Hiram Graves and Martin Pierce should be added to this list, despite their lack of identification with carpentry in Westford after 1850, because they represent part of the small group of known pre-1850 carpenters.

26. Jackson's "Medical Daybook" (NYSHA) covers the years 1833 to 1837. The other workers whom he names as performing some kind of woodworking are George Snyder, Alexander Lloyd, James Hubbard, Christopher Snyder, Thomas Webster, and Ezra Thurber. Except for Thurber, who clapboarded a shed for Jackson and who in any case is otherwise known as a carpenter in Westford, none of these men performed very skilled labor; mostly they scored timber (whereas Andrew Snyder *hewed* it), did fence work, and the like. Hubbard and Christopher Snyder also did various kinds of day labor for Jackson: butchering, reaping, sawing wood, pressing cider, drawing timber, and so forth.

27. The birthplaces of several men credited to various locales are not certain; the biographical sketches in appendix 2 specify these doubtful cases.

28. Two exceptions are Richardson Babcock and Erasmus Snyder, who went to Albany and Cherry Valley, respectively.

29. Had he survived, Henry well might have come back to Westford, as did his nephew William (no. 61) once the war was over, even though, also like William (who left and came back and left again before dying in Westford), he well might have left the township again. His brother Warren may have come back to Westford briefly after the war—he had a wife and child there in 1865, before he was discharged—but by 1870 he was gone.

30. It is not surprising in part because the problem here is somewhat circular: that is, if it were possible to track these men for longer periods of time, it is likely that more would be known about their origins, and vice versa.

31. Elias Loomis, with revisions by Elisha S. Loomis, *Descendants of James Loomis in America and His Antecedents in the Old World* (1875, 1908; rpt., n.p.: 1981), pp. 439, 550; plus local census records.

32. In fact, Emily, who at age thirty in 1855 reported that she had passed only twenty years in Westford township, may well have left the immediate vicinity even before her marriage in 1847. Her parents, aged fifty-six and fifty-two in 1855, had spent fifty years there by 1855, and all her siblings still resident in the parents' household then (there were five, ranging in age from eleven to twenty-six) had lived in Westford all their lives. Since Emily had joined the Westford church in 1838 and furthermore seems to have been counted in her

parents' household in the 1840 census, exactly when she could have been absent from Westford—say, as a domestic living in a household just across the boundary, a likely employment for a young teenage girl then—is not clear. Assuming that she and Loomis spent all their married life prior to 1855 outside Westford, she must have spent at least two years before 1847 outside the township on her own. Perhaps it was in Middlefield that she met Loomis.

33. Perhaps at some point the Loomises spent some time in nearby Maryland. The genealogy cited in note 31 says that Ida Loomis, the couple's second child, was born on September 4, 1850, in that township. If this is accurate (and in a number of details the listing for Charles Loomis is quite inaccurate) and the Loomises then were living in Maryland, perhaps there was even less stability in their lives than otherwise seems to have been the case.

34. The Loomis genealogy lists an earlier child, Sarah M., said to have been born in 1843, four years before the couple was married. She has proved to be otherwise untraceable.

35. Probably a Westford native, Babcock had circulated widely around this point, working in manufacturing and trades in Worcester in 1820, then living in Albany in 1830 before returning to Westford (and a manufacturing-and-trades listing there) in 1840; he was gone from Westford again in 1855, however, showing up nearby again in 1860, when he was a carpenter in Maryland township; in 1870 he was a farmer there, although an 1872 directory listing described him as a carpenter once more, just three years before his death in Maryland in 1875.

36. The following Minneapolis directories have been consulted: Merwin's (1867), p. 68; Tribune Co. (1869), p. 69, (1871–72), p. 44, (1873–74), p. 174; Campbell's (1874), p. 186, (1875), p. 198, (1876), p. 157, (1877–78), p. 168, (1878–79), p. 157, (1879–80), p. 142, (1881–82), p. 254.

37. The *Schenevus Valley News* printed Sarah Jane Groff's obituary on September 8, 1868; the *Minneapolis Daily Tribune* did so on August 13, 1868.

CONCLUSION: SKILL AND RESPONSIBILITY

1. Wendell Berry, *The Unsettling of America: Culture and Agriculture* (New York: Avon Books, 1978), p. 91.

2. Here the distinctions among seventeenth-century northeastern economic cultures made by Stephen Innes in *Labor in a New Land: Economy and Society in Seventeenth-Century Springfield* (Princeton: Princeton University Press, 1983) seem to me particularly worth attention. Innes proposes that the classic community taken by several recent scholars as the model of early New England life in fact represents only one type of settlement, the inland agricultural village

largely detached from the market. He puts Springfield forward as the countertype; it represents a group of river settlements that were from the start the sites of an aggressively commercial agriculture (and, in this case, fur trading). With modifications owing to the changed economic order of the nineteenth century, Westford might be thought of as continuing the tradition of the inland settlements.

3. Donald Worster, *Nature's Economy: A History of Ecological Ideas* (New York: Cambridge University Press, 1985), p. 227.

4. Bentley's account book, referred to earlier, is in the NYSHA collections in Cooperstown.

Index